quantum physics
a beginner's guide

"*Rae has done an impressive job . . . Any reader who is prepared to put in a little effort will come away from this book with not only an understanding of the basics of some important practical applications of the theory but also some appreciation of why its conceptual foundations are still the subject of such spirited debate.*"

Professor Anthony Leggett, winner of
the 2003 Nobel Prize for Physics

From anarchism to artificial intelligence and genetics to global terrorism, **BEGINNERS GUIDES** equip readers with the tools to fully understand the most challenging and important debates of our age. Written by experts in a clear and accessible style, books in this series are substantial enough to be thorough but compact enough to be read by anyone wanting to know more about the world they live in.

anarchism
ruth kinna

anti-capitalism
simon tormey

artificial intelligence
blay whitby

biodiversity
john spicer

bioterror & biowarfare
malcolm dando

the brain
a. al-chalabi, m. r. turner
& r. s. delamont

christianity
keith ward

cloning
aaron d. levine

criminal psychology
ray bull *et al.*

democracy
david beetham

energy
vaclav smil

evolution
burton s. guttman

evolutionary psychology
r. dunbar, l.barrett &
j. lycett

fair trade
jacqueline decarlo

genetics
a. griffiths, b.guttman,
d. suzuki & t. cullis

global terrorism
leonard weinberg

hinduism
klaus k. klostermaier

life in the universe
lewis dartnell

the mafia & organized crime
james o. finckenauer

NATO
jennifer medcalf

the palestine–israeli conflict
dan cohn-sherbok &
dawoud el-alami

philosophy of mind
edward feser

postmodernism
kevin hart

quantum physics
alastair i. m. rae

religion
martin forward

the small arms trade
m. schroeder, r. stohl
& d. smith

FORTHCOMING:

animal behaviour
beat generation
bioethics
british politics
censorship
christianity
climate change
conspiracy theories
crimes against humanity
engineering
ethics
existentialism

extrasolar planets
feminist theory
forensic science
galaxies
gender & sexuality
globalization
hinduism
human rights
humanism
immigration
indigenous peoples

modern slavery
oil
philosophy of religion
political philosophy
racism
radical philosophy
renaissance art
romanticism
socialism
time
volcanoes

quantum physics

a beginner's guide

alastair i.m. rae

ONEWORLD
OXFORD

quantum physics: a beginner's guide

Oneworld Publications
185 Banbury Road
Oxford OX2 7AR
England
www.oneworld-publications.com

First published 2005
Reprinted 2006, 2007

ISBN-13: 978–1–85168–369–7

Cover design by Two Associates
Typeset by SNP Best-set Typesetter Ltd., Hong Kong
Printed and bound by Biddles Ltd., King s Lynn

Figures 2.1b, 4.4a and 5.9 © istockphoto

To Amelia and Alex

contents

preface

The year 2005 is the 'World Year of Physics'. It marks the centenary of the publication of three papers by Albert Einstein during a few months in 1905. The most famous of these is probably the third, which set out the theory of relativity, while the second paper provided definitive evidence for the (then controversial) idea that matter was composed of atoms. Both had a profound effect on the development of physics during the rest of the twentieth century and beyond, but it is Einstein's first paper that led to quantum physics.

In this paper, Einstein showed how some recent experiments demonstrated that the energy in a beam of light travelled in packets known as 'quanta' (singular: 'quantum'), despite the fact that in many situations light is known to behave as a wave. This apparent contradiction was to lead to the idea of 'wave–particle duality' and eventually to the puzzle of Schrödinger's famous (or notorious) cat. This book aims to introduce the reader to a selection of the successes and triumphs of quantum physics; some of these lie in explanations of the behaviour of matter on the atomic and smaller scales, but the main focus is on the manifestation of quantum physics in everyday phenomena. It is not always realized that much of our modern technology has an explicitly quantum basis. This applies not only to the inner workings of the silicon chips that power our computers, but also to the fact that electricity can be conducted along metal wires and not through insulators. For many years now, there has been considerable concern about the effect of our technology on the environment and, in particular, how emission of carbon dioxide into the Earth's atmosphere is leading to global warming; this 'greenhouse effect' is also a manifestation of quantum physics, as are some of the green technologies being developed to counteract it. These phenomena are discussed here, as are the application of quantum physics to what is known as 'superconductivity' and to information technology. We address some of the more philosophical aspects of the subject towards the end of the book.

Quantum physics has acquired a reputation as a subject of great complexity and difficulty; it is thought to require considerable intellectual effort and, in particular, a mastery of higher mathematics. However, quantum physics need not be 'rocket science'. It is possible to use the idea of wave–particle duality to understand many important quantum phenomena without much, or any, mathematics. Accordingly, the main text contains practically no mathematics, although it is complemented by 'mathematical boxes' that flesh out some of the arguments. These employ only the basic mathematics many readers will have met at school, and the reader can choose to omit them without missing the main strands of the argument. On the other hand, the aim of this book is to lead readers to an *understanding* of quantum physics, rather than simply impressing them with its sometimes dramatic results. To this end, considerable use is made of diagrams and the reader would be well advised to study these carefully along with the text. Inevitably, technical terms are introduced from time to time and a glossary of these will be found towards the end of the volume. Some readers may already have some expertise in physics and will no doubt notice various simplifications of the arguments they have been used to. Such simplifications are inevitable in a treatment at this level, but I hope and believe that they have not led to the use of any incorrect models or arguments.

I should like to thank my former students and colleagues at the University of Birmingham, where I taught physics for over thirty years, for giving me the opportunity to widen and deepen my knowledge of the subject. Victoria Roddam and others at Oneworld Publications have shown considerable patience, while applying the pressure needed to ensure the manuscript was delivered, if not in time, then not too late. Thanks are also due to Ann and the rest of my family for their patience and tolerance. Finally, I of course take responsibility for any errors and inaccuracies.

Alastair I. M. Rae

quantum physics is not rocket science

'Rocket science' has become a byword in recent times for something really difficult. Rocket scientists require a detailed knowledge of the properties of the materials used in the construction of spacecraft; they have to understand the potential and danger of the fuels used to power the rockets and they need a detailed understanding of how planets and satellites move under the influence of gravity. Quantum physics has a similar reputation for difficulty, and a detailed understanding of the behaviour of many quantum phenomena certainly presents a considerable challenge – even to many highly trained physicists. The greatest minds in the physics community are probably those working on the unresolved problem of how quantum physics can be applied to the extremely powerful forces of gravity that are believed to exist inside black holes, and which played a vital part in the early evolution of our universe. However, the fundamental ideas of quantum physics are really not rocket science: their challenge is more to do with their unfamiliarity than their intrinsic difficulty. We have to abandon some of the ideas of how the world works that we have all acquired from our observation and experience, but once we have done so, replacing them with the new concepts required to understand quantum physics is more an exercise for the imagination than the intellect. Moreover, it is quite possible to understand how the principles of quantum mechanics underlie many everyday phenomena, without using the complex mathematical analysis needed for a full professional treatment.

1

The conceptual basis of quantum physics is strange and unfamiliar, and its interpretation is still controversial. However, we shall postpone most of our discussion of this to the last chapter,[1] because the main aim of this book is to understand how quantum physics explains many natural phenomena; these include the behaviour of matter at the very small scale of atoms and the like, but also many of the phenomena we are familiar with in the modern world. We shall develop the basic principles of quantum physics in Chapter 2, where we will find that the fundamental particles of matter are not like everyday objects, such as footballs or grains of sand, but can in some situations behave as if they were waves. We shall find that this 'wave–particle duality' plays an essential role in determining the structure and properties of atoms and the 'subatomic' world that lies inside them.

Chapter 3 begins our discussion of how the principles of quantum physics underlie important and familiar aspects of modern life. Called 'Power from the Quantum', this chapter explains how quantum physics is basic to many of the methods used to generate power for modern society. We shall also find that the 'greenhouse effect', which plays an important role in controlling the temperature and therefore the environment of our planet, is fundamentally quantum in nature. Much of our modern technology contributes to the greenhouse effect, leading to the problems of global warming, but quantum physics also plays a part in the physics of some of the 'green' technologies being developed to counter it.

In Chapter 4, we shall see how wave–particle duality features in some large-scale phenomena; for example, quantum physics explains why some materials are metals that can conduct electricity, while others are 'insulators' that completely obstruct such current flow. Chapter 5 discusses the physics of 'semiconductors' whose properties lie between those of metals and insulators. We shall find out how quantum physics plays an essential role in these materials, which have been exploited to construct the silicon chip. This device is the basis of modern electronics, which, in turn, underlies the information and communication technology that plays such an important role in the modern world.

In Chapter 6 we shall turn to the phenomenon of 'superconductivity', where quantum properties are manifested in a particularly dramatic manner: the large-scale nature of the quantum phenomena in this case produces materials whose resistance to the flow of electric current vanishes completely. Another intrinsically quantum phenomenon relates to recently developed techniques for processing information and we shall discuss some of these in Chapter 7. There we shall find that it is possible to use quantum physics to transmit information in a form that cannot be read by any unauthorized person. We shall also learn how it may one day be possible to build 'quantum computers' to perform some calculations many millions of times faster than can any present-day machine.

Chapter 8 returns to the problem of how the strange ideas of quantum physics can be interpreted and understood, and introduces some of the controversies that still rage in this field, while Chapter 9 aims to draw everything together and make some guesses about where the subject may be going.

As we see, much of this book relates to the effect of quantum physics on our everyday world: by this we mean phenomena where the quantum aspect is displayed at the level of the phenomenon we are discussing and not just hidden away in objects' quantum sub-structure. For example, although quantum physics is essential for understanding the internal structure of atoms, in many situations the atoms themselves obey the same physical laws as those governing the behaviour of everyday objects. Thus, in a gas the atoms move around and collide with the walls of the container and with each other as if they were very small balls. In contrast, when a few atoms join together to form molecules, their internal structure is determined by quantum laws, and these directly govern important properties such as their ability to absorb and re-emit radiation in the greenhouse effect (Chapter 3).

The present chapter sets out the background needed to understand the ideas I shall develop in later chapters. I begin by defining some basic ideas in mathematics and physics that were developed before the quantum era; I then give an account of some of the nineteenth-century discoveries, particularly about the nature of atoms, that revealed the need for the revolution in our thinking that became known as 'quantum physics'.

mathematics

To many people, mathematics presents a significant barrier to their understanding of science. Certainly, mathematics has been the language of physics for four hundred years and more, and it is difficult to make progress in understanding the physical world without it. Why is this the case? One reason is that the physical world appears to be largely governed by the laws of cause and effect (although these break down to some extent in the quantum context, as we shall see). Mathematics is commonly used to analyse such causal relationships: as a very simple example, the mathematical statement 'two plus two equals four' implies that if we take any two physical objects and combine them with any two others, we will end up with four objects. To be a little more sophisticated, if an apple falls from a tree, it will fall to the ground and we can use mathematics to calculate the time this will take, provided we know the initial height of the apple and the strength of the force of gravity acting on it. This exemplifies the importance of mathematics to science, because the latter aims to make predictions about the future behaviour of a physical system and to compare these with the results of measurement. Our belief in the reliability of the underlying theory is confirmed or refuted by the agreement, or lack of it, between prediction and measurement. To test this sensitively we have to represent the results of both our calculations and our measurements as numbers.

To illustrate this point further, consider the following example. Suppose it is night time and three people have developed theories about whether and when daylight will return. Alan says that according to his theory it will be daylight at some undefined time in the future; Bob says that daylight will return and night and day will follow in a regular pattern from then on; and Cathy has developed a mathematical theory which predicts that the sun will rise at 5.42 a.m. and day and night will then follow in a regular twenty-four-hour cycle, with the sun rising at predictable times each day. We then observe what happens. If the sun does rise at precisely the times Cathy predicted, all three theories will be verified, but we are likely to give hers considerably more credence. This is because if the sun had risen at some other time, Cathy's theory would have been disproved, or falsified, whereas Alan and Bob's would still have stood. As the philosopher Karl Popper pointed out, it is this potential for falsification that gives a physical theory its strength. Logically, we cannot know for certain that it is true, but our faith in it will be strengthened the more rigorous are the tests that it passes. To falsify Bob's theory, we would have to observe the sun rise, but at irregular times on different days, while Alan's theory would be falsified only if the sun never rose again. The stronger a theory is, the easier it is in principle to find that it is false, and the more likely we are to believe it if we fail to do so. In contrast, a theory that is completely incapable of being disproved is often described as 'metaphysical' or unscientific.

To develop a scientific theory that can make a precise prediction, such as the time the sun rises, we need to be able to measure and calculate quantities as accurately as we can, and this inevitably involves mathematics. Some of the results of quantum calculations are just like this and predict the values of measurable quantities to great accuracy. Often, however, our predictions are more like those of Bob: a pattern of behaviour is predicted rather than a precise number. This also involves mathematics, but we can often avoid the complexity needed to predict actual numbers, while still making predictions that are sufficiently testable to give us confidence in them if they pass such a test. We shall encounter several examples of the latter type in this book.

The amount of mathematics we need depends greatly on how complex and detailed is the system that we are studying. If we choose our examples appropriately we can often exemplify quite profound physical ideas with very simple calculations. Wherever possible, we limit the mathematics used in this book to arithmetic and simple algebra; however, our aim of describing real-world phenomena will sometimes lead us to discuss problems where a complete solution would require a higher level of mathematical analysis. In discussing these, we shall avoid mathematics as much as possible, but we shall be making extensive use of diagrams, which should be carefully studied along with the text. Moreover, we shall sometimes have to simply state results, hoping that the reader is prepared to take them on trust. A number of reasonably straightforward mathematical arguments relevant to our discussion are included in 'mathematical boxes' separate from the main text. These are not essential to our discussion, but readers who are more comfortable with mathematics may find them interesting and helpful. A first example of a mathematical box appears below as Mathematical Box 1.1.

MATHEMATICAL BOX 1.1

Although the mathematics used in this book is no more than most readers will have met at school, these are skills that are easily forgotten with lack of practice. At the risk of offending the more numerate reader, this box sets out some of the basic mathematical ideas that will be used.

A key concept is the *mathematical formula* or *equation*, such as

$$a = b + cd$$

In algebra, a letter represents some number, and two letters written together means that they are to be multiplied. Thus if, for example, b is 2, c is 3 and d is 5, c must equal $2 + 3 \times 5 = 2 + 15 = 17$.

Powers. If we multiply a number (say x) by itself we say that we have 'squared' it or raised it to power 2 and we write this as x^2. Three copies of the same number multiplied together (xxx) is x^3 and so on. We can also have negative powers and these are defined such that $x^{-1} = 1/x$, $x^{-2} = 1/x^2$ and so on.

An example of a formula used in physics is Einstein's famous equation:

$$E = mc^2$$

Here, E is energy, m is mass and c is the speed of light, so the physical significance of this equation is that the energy contained in an object equals its mass multiplied by the square of the speed of light. As an equation states that the right- and left-hand sides are always equal, if we perform the same operation on each side, the equality will still hold. So if we divide both sides of Einstein's equation by c^2, we get

$$E/c^2 = m \text{ or } m = E/c^2$$

where we note that the symbol / represents division and the equation is still true when we exchange its right- and left-hand sides.

classical physics

If quantum physics is not rocket science, we can also say that 'rocket science is not quantum physics'. This is because the motion of the sun and the planets as well as that of rockets and artificial satellites can be calculated with complete accuracy using the pre-quantum physics developed between two and three hundred years ago by Newton and others.[2] The need for quantum physics was not realized until the end of the nineteenth century, because in many familiar situations quantum effects are much too small to be significant. When we discuss quantum physics, we refer to this earlier body of knowledge as 'classical'. The word 'classical' is used in a number of scientific fields to mean something like 'what was known before the topic we are discussing became relevant', so in our context it refers to the body of scientific knowledge that preceded the quantum revolution. The early quantum physicists were familiar with the concepts of classical physics and used them where they could in developing the new ideas. We shall be following in their tracks, and will shortly discuss the main ideas of classical physics that will be needed in our later discussion.

units

When physical quantities are represented by numbers, we have to use a system of 'units'. For example, we might measure distance in miles, in which case the unit of distance would be the mile, and time in hours, when the unit of time would be the hour, and so on. The system of units used in all scientific work is known by the French name 'Systeme Internationale', or 'SI' for short. In this system the unit of distance is the metre (abbreviation 'm'), the unit of time is the second ('s'), mass is measured in units of kilograms ('kg') and electric charge in units of coulombs ('C').

The sizes of the fundamental units of mass, length and time were originally defined when the metric system was set up in the late eighteenth and early nineteenth century. Originally, the metre was defined as one ten millionth of the distance from the pole to the equator, along the meridian passing through Paris; the second as 1/86,400 of an average solar day; and the kilogram as the mass

of one thousandth of a cubic metre of pure water. These definitions gave rise to problems as our ability to measure the Earth's
dimensions and motion more accurately implied small changes in
these standard values. Towards the end of the nineteenth century,
the metre and kilogram were redefined as, respectively, the distance between two marks on a standard rod of platinum alloy, and
the mass of another particular piece of platinum; both these standards were kept securely in a standards laboratory near Paris and
'secondary standards', manufactured to be as similar to the originals as possible, were distributed to various national organiza-
tions. The definition of the second was modified in 1960 and
expressed in terms of the average length of the year. As atomic
measurements became more accurate, the fundamental units were
redefined again: the second is now defined as 9,192,631,770
periods of oscillation of the radiation emitted during a transition
between particular energy levels of the caesium atom,[3] while the
metre is defined as the distance travelled by light in a time equal
to 1/299,792,458 of a second. The advantage of these definitions is
that the standards can be independently reproduced anywhere on
Earth. However, no similar definition has yet been agreed for the
kilogram, and this is still referred to the primary standard held by
the French Bureau of Standards. The values of the standard
masses we use in our laboratories, kitchens and elsewhere have all
been derived by comparing their weights with standard weights,
which in turn have been compared with others, and so on until we
eventually reach the Paris standard.

The standard unit of charge is determined through the ampere,
which is the standard unit of current and is equivalent to one
coulomb per second. The ampere itself is defined as that current
required to produce a magnetic force of a particular size between
two parallel wires held one metre apart.

Other physical quantities are measured in units that are
derived from these four: thus, the speed of a moving object is
calculated by dividing the distance travelled by the time taken,
so unit speed corresponds to one metre divided by one second,
which is written as '$m\,s^{-1}$'. Note this notation, which is adapted
from that used to denote powers of numbers in mathematics (cf.
Mathematical Box 1.1). Sometimes a derived unit is given its own
name: thus, energy (to be discussed below) has the units of mass

times velocity squared so it is measured in units of $kg\,m^2\,s^{-2}$, but this unit is also known as the 'joule' (abbreviation 'J') after the nineteenth-century French scientist who discovered that heat was a form of energy.

In studying quantum physics, we often deal with quantities that are very small compared with those used in everyday life. To deal with very large or very small quantities, we often write them as numbers multiplied by powers of ten, according to the following convention: we interpret 10^n, where n is a positive whole number, as 1 followed by n zeros, so that 10^2 is equivalent to 100 and 10^6 to 1,000,000; while 10^{-n} means $n - 1$ zeros following a decimal point so that 10^{-1} is the same as 0.1, 10^{-5} represents 0.00001 and 10^{-10} means 0.0000000001. Some powers of ten have their own symbol: for example, 'milli' means one thousandth; so one millimetre (1 mm) is 10^{-3} m. Other such abbreviations will be explained as they come up. An example of a large number is the speed of light, whose value is $3.0 \times 10^8\,m\,s^{-1}$, while the fundamental quantum constant (known as 'Planck's constant' – see below) has the value $6.6 \times 10^{-34}\,J\,s$. Note that to avoid cluttering the text with long numbers, I have quoted the values of these constants to one place of decimals only; in general, I shall continue this practice throughout, but we should note that most fundamental constants are nowadays known to a precision of eight or nine places of decimals and important experiments have compared experimental measurements with theoretical predictions to this precision (for an example see Mathematical Box 2.7 in Chapter 2).

motion

A substantial part of physics, both classical and quantum, concerns objects in motion, and the simplest concept used here is that of speed. For an object moving at a steady speed, this is the distance (measured in metres) it travels in one second. If an object's speed varies, then its value at any given time is defined as the distance it would have travelled in one second if its speed had remained constant. This idea should be familiar to anyone who has travelled in a motorcar, although the units in this case are normally kilometres (or miles) per hour.

Closely related to the concept of speed is that of 'velocity'. In everyday speech these terms are synonymous, but in physics they are distinguished by the fact that velocity is a 'vector' quantity, which means that it has direction as well as magnitude. Thus, an object moving from left to right at a speed of $5\,\mathrm{m\,s^{-1}}$ has a positive velocity of $5\,\mathrm{m\,s^{-1}}$, but one moving at the same speed from right to left has a negative velocity of $-5\,\mathrm{m\,s^{-1}}$. When an object's velocity is changing, the rate at which it does so is known as acceleration. If, for example, an object's speed changes from $10\,\mathrm{m\,s^{-1}}$ to $11\,\mathrm{m\,s^{-1}}$ during a time of one second, the change in speed is $1\,\mathrm{m\,s^{-1}}$ so its acceleration is 'one metre per second per second' or $1\,\mathrm{m\,s^{-2}}$.

mass

Isaac Newton defined the mass of a body as 'the quantity of matter' it contains, which begs the question of what matter is or how its 'quantity' can be measured. The problem is that, though we can define some quantities in terms of more fundamental quantities (e.g. speed in terms of distance and time), some concepts are so fundamental that any such attempt leads to a circular definition like that just stated. To escape from this, we can define such quantities 'operationally', by which we mean that we describe what they do – i.e. how they operate – rather than what they are. In the case of mass, this can be done through the force an object experiences when exposed to gravity. Thus two bodies with the same mass will experience the same force when placed at the same point of the Earth's surface, and the masses of two bodies can be compared using a balance.[4]

energy

This is a concept we shall be frequently referring to in our later discussions. An example is the energy possessed by a moving body, which is known as 'kinetic energy'; this is calculated as one half of the mass of the body by the square of its speed – see Mathematical Box 1.2 – so its units are joules, equivalent to $\mathrm{kg\,m^2\,s^{-2}}$. Another important form of energy is 'potential energy', which is associated with the force acting on a body. An example is the potential energy associated with gravity, which increases in

MATHEMATICAL BOX 1.2

To express the concept of energy quantitatively, we first have to express the kinetic and potential energies as numbers that can be added to produce a number for the total energy. In the text, we define the kinetic energy of a moving object as one half of the product of the mass of the object with the square of its speed. If we represent the mass by the symbol m, the speed by v and the kinetic energy by K, we have

$$K = \frac{1}{2}mv^2$$

In the case of an object falling to the surface of the Earth its potential energy is defined as the product of the mass (m) of the object, its height (h) and a constant g, known as the 'acceleration due to gravity', which has a value close to $10\,\mathrm{ms^{-2}}$. Thus, calling the potential energy V,

$$V = mgh$$

The total energy, E, is then

$$E = K + V = \frac{1}{2}mv^2 + mgh$$

Suppose that our object has a mass of 1 kilogram and is released one metre above the floor. At this point it has zero kinetic energy (because it hasn't started moving yet) and a potential energy of 10 J. As it reaches the floor, the total energy is still 10 J (because it is conserved), but the potential energy is zero. The kinetic energy must therefore now be 10 J, which means that the object's speed is about $4.5\,\mathrm{ms^{-1}}$.

proportion to the distance an object is raised from the floor. Its value is calculated by multiplying the object's mass by its height and then by the acceleration due to gravity. The units of these three quantities are kg, m and $\mathrm{m\,s^{-2}}$, respectively, so the unit of potential energy is $\mathrm{kg\,m^2\,s^{-2}}$, which is the same as that of kinetic energy, as is to be expected because different forms of energy can be converted from one to the other.

An extremely important principle in both quantum and classical physics is that of 'conservation of energy'; which means that energy can never be created or destroyed. Energy can be converted from one form to another, but the total amount of energy always remains the same. We can illustrate this by considering one of the simplest examples of a physical process, an object falling under gravity. If we take any object and drop it, we find that it moves faster and faster as it drops to the ground. As it moves, its potential energy becomes less and its speed and therefore kinetic energy increase. At every point the total energy is the same.

Now consider what happens after the falling object lands on the Earth. Assuming it doesn't bounce, both its kinetic and potential energies have reduced to zero, so where has the energy gone? The answer is that it has been converted to heat, which has warmed up the Earth around it (see the section on temperature below). This is only a small effect in the case of everyday objects, but when large bodies fall the energy release can be enormous: for example, the collision of a meteorite with the Earth many million years ago is believed to have led to the extinction of the dinosaurs. Other examples of forms of energy are electrical energy (which we shall be returning to shortly), chemical energy, and mass energy as expressed in Einstein's famous equation, $E = mc^2$.

electric charge

There are two main sources of potential energy in classical physics. One is gravity, which we referred to above, while the other is electricity, sometimes associated with magnetism and called 'electromagnetism'. A fundamental concept in electricity is electrical charge and, like mass, it is a quantity that is not readily defined in terms of other more fundamental concepts, so we again resort to an operational definition. Two bodies carrying electrical charge exert a force on each other. If the charges have the same sign this force is repulsive and pushes the bodies away from each other, whereas if the signs are opposite it is attractive and pulls them together. In both cases, if the bodies were released they would gain kinetic energy, flying apart in the like-charge case or together if the charges are opposite. To ensure that energy is conserved, there must be a potential energy associated with the interaction between

MATHEMATICAL BOX 1.3

The mathematical expression for the potential energy of inter-action between two charges of magnitude q_1 and q_2, separated by a distance r is

$$V = kq_1q_2/r$$

Where k is a constant defined so that the energy is calculated in joules when charge is measured in coulombs and distance in metres. Its value is $9.0 \times 10^9 \, \text{J} \, \text{m} \, \text{C}^{-2}$. We see that as the charges come closer together so that r reduces, then V gets larger (i.e. more positive) if the charges have the same sign, whereas it gets smaller (i.e. becomes more negative) if the signs of q_1 and q_2 are opposite.

the charges, one that gets larger as the like charges come together or as the unlike charges separate. More detail is given in Mathematical Box 1.3.

electric fields

When two electric charges interact, the presence of one causes a force to act on the other and as a result both start moving, either away from each other if the charges have the same sign or towards each other if the signs are opposite. The question arises of how one charge can know that the other exists some distance away. To answer this, physicists postulate that an electric charge creates an 'electric field' throughout space, which in turn acts on another charge to produce the electrical force. Field is therefore another fundamental concept that is defined operationally – cf. our earlier definitions of mass and charge. Evidence to support this concept comes from experiments in which both charges are initially held fixed and one of them is then moved. It is found that the force on the other does not change straight away, but only after a time equivalent to that taken by light to travel the distance between the charges. This means that the field created by the moving particle takes time to respond, the parts of the field near the moving charge changing before those further away.

When charges move, not only does the electric field change, but another field, the 'magnetic field', is created. Familiar examples of this field are that created by a magnet or indeed by the Earth, which controls the direction of a compass needle. The coupled electric and magnetic fields created by moving charges propagate through space in the form of 'electromagnetic waves', one example of which is light waves. We shall return to this in more detail in Chapter 2.

momentum

The momentum of a moving body is defined as the product of its mass and its velocity, so a heavy object moving slowly can have the same momentum as a light body moving quickly. When two bodies collide, the total momentum of both stays the same so that momentum is 'conserved' just as in the case of energy discussed earlier. However, momentum is different from energy in an important respect, which is that it is a vector quantity (like velocity) that has direction as well as magnitude. When we drop a ball on the ground and it bounces upwards at about the same speed, its momentum changes sign so that the total momentum change equals twice its initial value. Given that momentum is conserved, this change must have come from somewhere and the answer to this is that it has been absorbed in the Earth, whose momentum changes by the same amount in the opposite direction. However, because the Earth is enormously more massive than the ball, the velocity change associated with this momentum change is extremely small and undetectable in practice. Another example of momentum conservation is a collision between two balls, such as on a snooker table as illustrated in Figure 1.1, where we see how the conservation of momentum involves direction as well as magnitude.

temperature

The significance of temperature to physics is that it is a measure of the energy associated with heat. As we shall discuss shortly, all

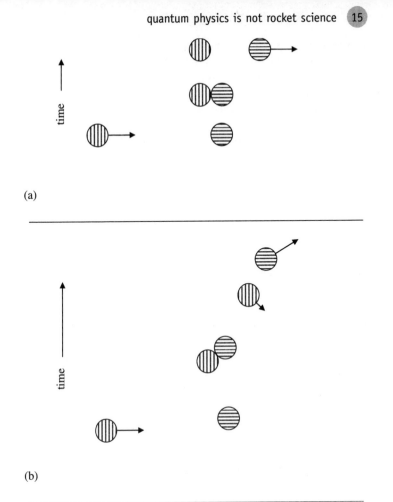

(a)

(b)

Figure 1.1 Snooker balls colliding. In (a) the left-hand ball approaches the stationary ball from the left (bottom line). They then collide (middle line) and the momentum is transferred from the left- to the right-hand ball, which moves away, leaving the left-hand ball stationary.

In (b), the collision is not head to head and both balls move away from the collision with the total momentum shared between them. Each particle now moves up or down at the same time as moving from left to right. The total momentum associated with the up-and-down motion is zero because one ball moves up while the other moves down and the total left-to-right momentum is the same as the left-hand one had initially. NB. the lengths and directions of the arrows indicate the particle velocities.

matter is composed of atoms. In a gas such as the air in a room, these are continually in motion and therefore possess kinetic energy. The higher the temperature of the gas, the higher is their average kinetic energy, and if we cool the gas to a lower temperature, the molecules move more slowly and the kinetic energy is less. If we were to continue this process, we should eventually reach a point where the molecules have stopped moving so that the kinetic energy and hence the temperature is zero. This point is known as the 'absolute zero of temperature' and corresponds to −273 degrees on the Celsius scale. The atoms and molecules in solids and liquids are also in thermal motion, though the details are rather different: in solids, for example, the atoms are held close to particular points, and vibrate around these. However, in every case this thermal motion reduces as the temperature is lowered and ceases as absolute zero is approached.[5] We use the concept of absolute zero to define an 'absolute scale' of temperature. In this scale, the degree of temperature has the same size as that on the Celsius scale, but the zero corresponds to absolute zero. Temperatures on this scale are known as 'absolute temperatures' or 'kelvins' (abbreviated as 'K') after the physicist Lord Kelvin, who made major contributions to this field. Thus, zero degrees absolute (i.e. 0 K) corresponds to −273°C, while a room temperature of 20°C is equivalent to 293 K, the boiling point of water (100°C) is 373 K and so on.

a first look at quantum objects

The need for fundamentally new ideas in physics emerged in the latter half of the nineteenth century when scientists found themselves unable to account for some of the phenomena that had recently been discovered. Some of these related to a detailed study of light and similar radiation, to which we shall return in the next chapter, while others arose from the study of matter and the realization that it is composed of 'atoms'.

the atom

Ever since the time of the ancient Greek philosophers there had been speculation that if matter were divided into smaller and smaller parts, a point would be reached where further subdivision was impossible. These ideas were developed in the nineteenth century, when it was realized that the properties of different chemical elements could be attributed to the fact that they were composed of atoms that were identical in the case of a particular element but differed from element to element. Thus a container of hydrogen gas is composed of only one type of atom (known as the hydrogen atom), a lump of carbon only another type (i.e. carbon atoms) and so on. By various means, such as studies of the detailed properties of gases, it became possible to estimate the size and mass of atoms. As expected, these are very small on the scale of everyday objects: the size of an atom is about 10^{-10} m and it weighs between about 10^{-27} kg in the case of hydrogen and 10^{-24} kg in the case of uranium (the heaviest naturally occurring element).

Although atoms are the smallest objects that carry the identity of a particular element, they have an internal structure, being constructed from a 'nucleus' and a number of 'electrons'.

the electron

Electrons are particles of matter that weigh much less than the atoms that contain them – the mass of an electron is a little less than 10^{-30} kg. They are 'point particles', which means that their size is zero – or at least too small to have been measured by any experiments conducted to date. All electrons carry an identical negative electric charge.

the nucleus

Nearly all the mass of the atom is concentrated in a 'nucleus' that is much smaller than the atom as a whole – typically 10^{-15} m in diameter or about 10^{-5} times the diameter of the atom. The nucleus carries a positive charge equal and opposite to the total charge carried by the electrons, so that the atom is uncharged or 'neutral' overall. It is known that the nucleus can be further divided into a number of positively charged particles known as 'protons' along with some uncharged particles known as 'neutrons'; the charge on the proton is positive, being equal and opposite to that on the electron. The masses of the neutron and proton are very similar (though not identical), both being about two thousand times the electron mass. Examples of nuclei are the hydrogen nucleus, which contains one proton and no neutrons; the nucleus of carbon, which contains six protons and six neutrons; and the uranium nucleus, which contains ninety-two protons and between 142 and 146 neutrons – see 'isotopes' below. When we want to refer to one of the particles making up the nucleus without specifying whether it is a proton or a neutron, we call it a 'nucleon'.

Nucleons are not point particles, like the electron, but have a structure of their own. They are each constructed from three point particles known as 'quarks'. Two kinds of quarks are found in the nucleus and these are known as the 'up' quark and the 'down' quark, though no physical significance should be attached to these labels. Up and down quarks carry positive charges of value $\frac{2}{3}$ and $-\frac{1}{3}$ respectively of the total charge on a proton, which contains two up quarks and one down quark. The neutron is constructed from one up quark and two down quarks, which is consistent with its zero overall charge. The quarks inside a neutron or proton are bound together very tightly so that the nucleons can be treated as single particles in nearly all circumstances. The neutrons and protons interact less strongly, but still much more strongly than the electrons interact with them, which means that to a very good approximation a nucleus can also be treated as a single particle, and its internal structure ignored when we are considering the structure of the atom. All this is illustrated in Figure 1.2, using the helium atom as an example.

isotopes

Most of the properties of atoms are derived from the electrons and the number of negatively charged electrons equals the number of positively charged protons in the nucleus. However, as noted above, the nucleus also contains a number of uncharged neutrons, which add to the mass of the nucleus but otherwise do not greatly affect the properties of the atom. If two or more atoms have the same number of electrons (and therefore protons) but different numbers of neutrons, they are known as 'isotopes'. An example is 'deuterium', whose nucleus contains one proton and

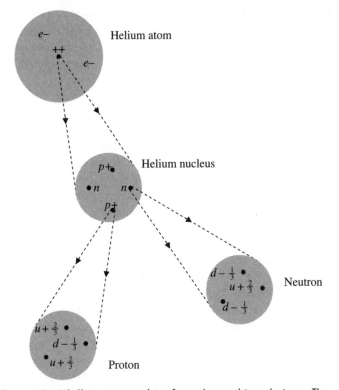

Figure 1.2 A helium atom consists of a nucleus and two electrons. The nucleus contains two protons and two neutrons, while the proton and neutron are composed of two up quarks with one down quark and two down quarks with one up quark, respectively. NB. no significance should be attached to the indicated positions of the electrons in the atom, the nucleons in the nucleus or the quarks in the nucleons.

one neutron and which is therefore an isotope of hydrogen; in naturally occurring hydrogen, about one atom in every ten thousand is deuterium.

The number of isotopes varies from element to element and is larger for heavier elements – i.e. those with a greater number of nucleons. The heaviest naturally occurring element is uranium, which has nineteen isotopes, all of which have 92 protons. The most common of these is U^{238}, which contains 146 neutrons, while the isotope involved in nuclear fission (see Chapter 3) is U^{235} with 143 neutrons. Note the notation where the superscript number is the total number of nucleons.

atomic structure

So far, we have seen that an atom consists of a very small positively charged nucleus surrounded by a number of electrons. The simplest atom is that of hydrogen, with one electron, and the biggest naturally occurring atom is that of uranium, which contains ninety-two electrons. Remembering that the nucleus is very small and that the dimensions of the electron are effectively zero, it is clear that much of the volume occupied by the atom must be empty space. This means that the electrons must stay some distance from the nucleus, despite the fact that there is an electrical attraction between each negatively charged electron and the positively charged nucleus. Why then does the electron not fall into the nucleus? One idea, suggested early in the development of the subject, is that the electrons are in orbit round the nucleus rather like the planets orbiting the sun in the solar system. However, a big difference between satellite orbits in a gravitational field and those where the orbiting particles are charged is that orbiting charges are known to lose energy by emitting electromagnetic radiation such as light. To conserve energy they should move nearer the nucleus where the potential energy is lower, and calculations show that this should lead to the electron collapsing into the nucleus within a small fraction of a second. However, for the atom to have its known size, this cannot and does not happen. No model based on classical physics is able to account for this observed property of atoms, and a new physics, quantum physics, is required.

A simple property of atoms that is inexplicable from a classical viewpoint is that all the atoms associated with a particular element are identical. Provided it contains the right number of electrons and a nucleus carrying a compensating positive charge, the atom will have all the properties associated with the element. Thus a hydrogen atom contains one electron and all hydrogen atoms are identical. To see why this is surprising classically, think again of a classical orbiting problem. If we put a satellite into orbit around the Earth, then, provided we do the rocket science properly, it can be at any distance from the Earth that we like. But all hydrogen atoms are the same size, which not only means that their electrons must be held at some distance from the nucleus, but also implies that this distance is the same for all hydrogen atoms at all times (unless, as we discuss below, an atom is deliberately 'excited'). Once again we see that the atom has properties that are not explicable using the concepts of classical physics.

To pursue this point further, consider what we might do to an atom to change its size. As moving the electron further from the nucleus increases its electrical potential energy, which has to come from somewhere, we would have to inject energy into the atom. Without going too far into the practical details, this can be achieved by passing an electrical discharge through a gas composed of the atoms. If we do this, we find that energy is indeed absorbed and then re-emitted in the form of light or other forms of electromagnetic radiation: we see this occurring whenever we switch on a fluorescent light. It seems that when we excite the atom in this way, it returns to its initial state by emitting radiation, rather as we predicted in the case of a charge in a classical orbit. However, there are two important differences in the atomic case. The first, discussed above, is that the final configuration of the atom corresponds to the electron being some distance from the nucleus and this state is always the same for all atoms of the same type. The second difference is related to the nature of the radiation emitted. Radiation has the form of electromagnetic waves, which will be discussed in more detail in the next chapter; for the moment, we need only know that such a wave has a characteristic wavelength corresponding to the colour of the light. Classically, a spiralling charge should emit light of all colours, but when the light emitted from an atomic discharge is examined, it is found to contain only certain colours that correspond to particular wavelengths. In the case of hydrogen, these form a reasonably simple pattern and it was one of the major early triumphs of quantum physics that it was able to predict this quite precisely. One of the new ideas that this is based on is the concept that the possible values of the energy of an atom are restricted to certain 'quantized' values, which include a lowest value or 'ground state' in which the electron remains some distance from the nucleus. When the atom absorbs energy, it can do so only if the energy ends up with one of the other allowed values, in which case the atom is said to be in an 'excited state', with the electron further from the nucleus than it is in the ground state. Following this, it returns to its ground state emitting radiation whose wavelength is determined by the difference in energy between the initial and final states.

None of the above phenomena can be accounted for using classical physics, but they can all be understood using the new quantum physics, as we shall see in the next chapter.

summary

In this introductory chapter, I have discussed a number of concepts that will be extensively used in later chapters:

- velocity, which is speed in a given direction;
- mass, which is the quantity of matter in a body;
- energy, which comes in a number of forms, including kinetic and potential energy;
- electrical charge and field, which relate to the energies of interaction of charged bodies;
- momentum, which is the velocity of a moving body multiplied by its mass;
- temperature, which is a measure of the energy associated with random motion of atoms and molecules.

We have seen that all matter is composed of atoms, which in turn consist of a nucleus surrounded by a number of electrons. Some of the properties of atoms cannot be understood using classical physics. In particular:

- All atoms of a given element are identical.
- Although attracted by the nucleus, the electrons do not collapse into it, but are held some distance away from it.
- The energy of an atom is 'quantized', meaning that its value always equals one of a set of discrete possibilities.

notes

1. I have also discussed the conceptual basis of quantum physics in *Quantum Physics: Illusion or Reality*, 2nd edn. Cambridge, Cambridge University Press, 2004.
2. However, when rocket scientists develop new construction materials or fuels, for example, they use and apply concepts and principles that rely explicitly or implicitly on the underlying quantum physics – see Chapter 3.
3. Energy levels and transitions between them will be discussed later in this chapter and in the next.
4. The reader may have been taught about the importance of distinguishing between 'mass' and 'weight': the latter is defined as the force on the object at the Earth's surface and this varies as we move to different parts of the globe. However, provided we make the measurements in the same place, we can validly compare masses through their weights.
5. It is never possible to quite reach absolute zero, but we can get extremely close to it. Temperatures as low as 10^{-9} K have been created in some specialist laboratories.

waves and particles

Many people have heard that 'wave–particle duality' is an important feature of quantum physics. In this chapter, we shall try to understand what this means and how it helps us to understand a range of physical phenomena, including the question of atomic structure that I introduced at the end of the previous chapter. We shall find that at the quantum level the outcomes of many physical processes are not precisely determined and the best we can do is to predict the likelihood or 'probability' of various possible events. We will find that something called the 'wave function' plays an important role in determining these probabilities: for example, its strength, or intensity, at any point represents the probability that we would detect a particle at or near that point. To make progress, we have to know something about the wave function appropriate to the physical situation we are considering. Professional quantum physicists calculate it by solving a rather complex mathematical equation, known as the Schrödinger equation (after the German physicist Erwin Schrödinger who discovered this equation in the 1920s); however, we will find that we can get quite a long way without doing this. Instead, we shall build up a picture based on some basic properties of waves, and we begin with a discussion of these as they feature in classical physics.

All of us have some familiarity with waves. Those who have lived near or visited the seacoast or have travelled on a ship will be aware of ocean waves (Figure 2.1[a]). They can be very large, exerting violent effects on ships, and they provide entertainment for surfers when they roll on to a beach. However, for our

(a)

(b)

Figure 2.1 (a) Waves on Bondi Beach. (b) Ripples on a pond.

purposes, it will be more useful to think of the more gentle waves or ripples that result when an object, such as a stone, is dropped into a calm pond (Figure 2.1[b]). These cause the surface of the water to move up and down so as to form a pattern in which ripples spread out from the point where the stone was dropped. Figure 2.2 shows a profile of such a wave, illustrating how it changes in time at different places. At any particular point in

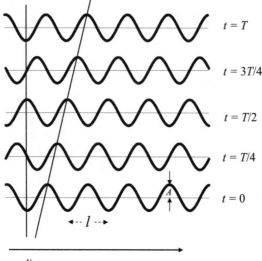

$t = T$

$t = 3T/4$

$t = T/2$

$t = T/4$

$t = 0$

A

$\leftarrow l \rightarrow$

distance

Figure 2.2   A water wave consists of a series of ripples containing peaks and troughs. At any instant, the distance between successive crests (or troughs) is known as the wavelength l. The maximum height of the wave is its amplitude A. The figure shows the form of the wave at a number of times, with the earliest ($t = 0$) at the foot. If we follow the vertical thin line, we see that the water surface has oscillated and returned to its original position after a time T, known as the period of the wave. The sloping thin line shows that during this time a particular crest has moved a distance l. It follows that the wave pattern moves at a speed c equal to l/T – see Mathematical Box 2.1.

space, the water surface oscillates up and down in a regular manner. The height of the ripple is known as the 'amplitude' of the wave, and the time taken for a complete oscillation is known as the 'period'. Often it is useful to refer to the 'frequency' of the wave, which is the number of times per second it moves through a complete cycle of oscillation. At any instant in time, the shape of the wave repeats in space, and the repeat distance is known as the 'wavelength'. During a time corresponding to one period, the pattern moves along a distance equal to the wavelength, which means that the wave moves at a speed corresponding to one wavelength per period (see Mathematical Box 2.1).

MATHEMATICAL BOX 2.1

We denote the wavelength of the wave by l and the period by T. It follows that the frequency of the wave is

$$f = 1/T$$

and the speed is given by

$$c = l/T$$

travelling waves and standing waves

Waves such as those illustrated in Figure 2.2 are what are called 'travelling waves' because they 'travel' in space. In the example shown, the motion is from left to right, but it could also have been from right to left. Indeed, we see from Figure 2.1(b) that the ripples spreading out from a stone dropped in water spread out in all directions.

As well as travelling waves, we shall need to know about 'standing waves'. An example is shown in Figure 2.3, where we see that the wave has a similar shape to that discussed earlier and the water again oscillates up and down, but now the wave does not move along, but stays in the same place – hence its name. A standing wave typically occurs when it is confined within a 'cavity' enclosed by two boundaries. If a travelling wave is set up, it is reflected at one of the boundaries and moves back in the opposite direction. When the waves travelling in the two directions are combined, the net result is the standing wave illustrated in Figure 2.3. In many cases, the walls of the cavity are such that the wave is unable to penetrate them and this results in the wave amplitude being equal to zero at the cavity boundaries.[1] This means that only standing waves of particular wavelengths are able to fit into the cavity – because, for the wave to be zero at both boundaries, its wavelength must be just the right length for a whole number of peaks or troughs to fit into the cavity. This is discussed in more detail in Mathematical Box 2.2.

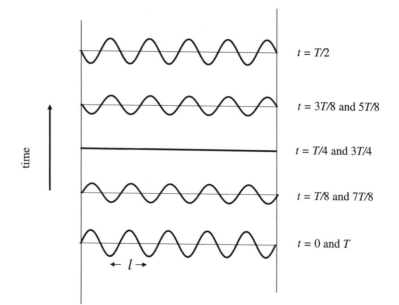

$t = T/2$

$t = 3T/8$ and $5T/8$

$t = T/4$ and $3T/4$

$t = T/8$ and $7T/8$

$t = 0$ and T

time

$\leftarrow l \rightarrow$

Figure 2.3 Standing waves occur when a wave is confined to a region in space. The wave moves up and down in time, but not in space.

MATHEMATICAL BOX 2.2

Referring to Figure 2.3, we see that if a standing wave has zero amplitude at the ends of a cavity of length L, then a whole number of half wavelengths must exactly fit into the distance L. Thus

$$L = \frac{1}{2} n l_n \text{ so that } l_n = 2L/n$$

where n is a whole number and l_n is one of the allowed wavelengths. The subscript n in l_n is simply a label used to distinguish the wavelengths belonging to the different standing waves. Thus

$$l_1 = 2L, l_2 = L, l_3 = 2L/3 \text{ and so on}$$

As the frequency of a wave is related to the wavelength, this must also be constrained to a set of particular values, given by

$$f_n = c/l_n = nc/2L$$

This principle underlies the operation of many musical instruments. For example, the note emitted by a violin or guitar is determined by the frequencies of the allowed standing waves on the string, which in turn are controlled by the length of string the player sets in oscillation. To change the pitch of the note, the player presses the string down at a different point so as to change the length of the vibrating part of the string.[2] Standing waves play a similar role in all musical instruments: woodwind and brass set up standing waves in confined volumes of air, while the sound emitted by drums comes from the standing waves set up in the drum skins. The types of sound produced by different musical instruments are very different – because the notes produced have different 'harmonic contents'. By this we mean that the vibration is not a simple 'pure' note corresponding to one of the allowed frequencies, but is constructed from a combination of standing waves, all of whose frequencies are multiples of the lowest or 'fundamental' frequency.

However, if the standing waves were the whole story, the sound would never reach our ears. For the sound to be transmitted to the listener, the vibrations of the instrument must generate travelling waves in the air, which carry the sound to the listener. In a violin, for example, the body of the instrument oscillates in sympathy with the string and generates a travelling wave that radiates out to the audience. Much of the science (or art) of designing musical instruments consists of ensuring that the frequencies of the notes determined by the allowed wavelengths of the standing waves are reproduced in the emitted travelling waves. A full understanding of the behaviour of musical instruments and the way they transmit sound to a listener is a major topic in itself, which we do not need to go into any further here. Interested readers should consult a book on the physics of music.

light waves

Another commonly encountered wavelike phenomenon is 'electromagnetic radiation', exemplified in the radio waves that bring signals to our radios and televisions and in light. These waves have different frequencies and wavelengths: for example, typical FM radio signals have a wavelength of 3 m, whereas the wavelength of light depends on its colour, being about 4×10^{-8} m for blue light and 7×10^{-8} m for red light; other colours have wavelengths between these values.

Light waves are different from water waves and sound waves in that there is nothing corresponding to the vibrating medium (i.e. the water, string or air) in the examples discussed earlier. Indeed, light waves are capable of travelling through empty space, as is obvious from the fact that we can see the light emitted by the sun and stars. This property of light waves presented a major problem to scientists in the eighteenth and nineteenth centuries. Some concluded that space is not actually empty, but filled with an otherwise undetectable substance known as 'aether' which was thought to support the oscillation of light waves. However, this hypothesis ran into trouble when it was realized that the properties required to support the very high frequencies typical of light could not be reconciled with the fact that the aether offers no resistance to the movement of objects (such as the Earth in its orbit) through it.

It was James Clerk Maxwell who around 1860 showed that the aether postulate was unnecessary. At that time, the physics of electricity and magnetism was being developed and Maxwell was able to show that it was all contained in a set of equations (now known as 'Maxwell's equations'). He also showed that one type of solution to these equations corresponds to the existence of waves that consist of oscillating electric and magnetic fields that can travel through empty space without requiring a medium. The speed these 'electromagnetic' waves travel at is determined by the fundamental constants of electricity and magnetism, and when this speed was calculated, it was found to be identical to the measured speed of light. This led directly to the idea that light is an electromagnetic wave and it is now known that this model also applies to a range of other phenomena, including radio waves, infrared radiation (heat) and X-rays.

interference

Direct evidence that a phenomenon, such as light, is a wave is obtained from studying 'interference'. Interference is commonly

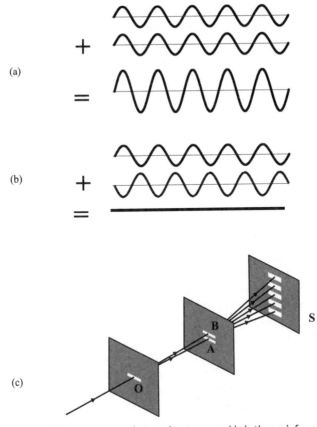

Figure 2.4 When two waves that are in step are added, they reinforce each other as in (a), but if they are exactly out of step they cancel each other out as in (b). Young's experiment is illustrated in (c). Light waves reaching a point on the screen S can have travelled through the slit O and then via one or other of the two slits A and B, so when they combine they have travelled different distances and an interference pattern consisting of a series of light and dark bands is observed on the screens.

encountered when two waves of the same wavelength are added together. Referring to Figure 2.4(a), we see that if the two waves are in step (the technical term is 'in phase') they add together to produce a combined wave that has twice the amplitude of either of the originals. If, on the other hand, they are exactly out of step (in 'antiphase') they cancel each other out (Figure 2.4[b]). In intermediate situations the waves partially cancel and the combined amplitude has a value between these extremes. Interference is crucial evidence for the wave properties of light and no other classical model can account for this effect. Suppose, for example, that we instead had two streams of classical particles: the total number of particles would always equal the sum of the numbers in the two beams and they would never be able to cancel each other out in the way that waves can.

The first person to observe and explain interference was Thomas Young, who around 1800 performed an experiment like that illustrated in Figure 2.4(c). Light passes through a narrow slit labelled O, after which it encounters a screen containing two slits, A and B, and finally reaches a third screen, S, where it is observed. The light reaching the last screen can have travelled by one of two routes – either by A or by B. However, the distances travelled by the light waves following these two paths are not equal, so they do not generally arrive at the screen in step with each other. It follows from the discussion in the previous paragraph that at some points on S the weaves will reinforce each other, while at others they will cancel; as a result, a pattern consisting of a series of light and dark bands is observed on the screen.

Despite all this, we shall soon see that there is evidence that light does exhibit particle properties in some circumstances and a fuller understanding of the quantum nature of light will introduce us to 'wave–particle duality'.

light quanta

Around the end of the nineteenth century and the beginning of the twentieth, evidence began to emerge that indicated that describing light as a wave is not sufficient to account for all its observed properties. Two particular areas of study were central in this. The first concerns the properties of the heat radiation emitted by hot objects. At reasonably high temperatures, this heat radiation becomes visible and we describe the object as 'red hot' or, at even higher temperatures, 'giving off a white heat'. We note that red corresponds to the longest wavelength in the optical spectrum, so it appears that light of long wavelength can be generated more easily (i.e. at a lower temperature) than that of shorter wavelength; indeed, heat radiation of longer wavelengths is commonly known as 'infrared'. Following the emergence of Maxwell's theory of electromagnetic radiation and progress in the understanding of heat (a topic to which Maxwell also made major contributions), physicists tried to understand these properties of heat radiation. It was known by then that temperature is related to energy: the hotter an object is, the more heat energy it contains. Also, Maxwell's theory predicted that the energy of an electromagnetic wave should depend only on its amplitude and, in particular, should be independent of its wavelength. One might therefore expect that a hot body would radiate at all wavelengths, the radiation becoming brighter, but not changing colour, as the temperature rises. In fact, detailed calculations showed that because the number of possible waves of a given wavelength increases as the wavelength reduces, shorter wavelength heat radiation should actually be brighter than that with long wavelengths, but again this should be the same at all temperatures. If this were true, all objects should appear violet in colour, their overall brightness being low at low temperatures and high at high temperatures, which of course is not what we observe. This discrepancy between theory and observation was known as the 'ultraviolet catastrophe'.

In an attempt to resolve the ultraviolet catastrophe, the physicist Max Planck proposed in 1900 that the conventional laws of electromagnetism should be modified so that electromagnetic wave energy always appeared in packets containing a fixed amount of energy. He also postulated that the energy contained in any one of these packets is determined by the frequency of the wave, being greater for higher frequencies (i.e. shorter wavelengths). More precisely, he postulated that each carried an amount of energy equal to the frequency multiplied by a constant number that is now known as 'Planck's constant' and believed to be a fundamental constant of nature; its value is about 6.6×10^{-23} J s. Such a packet of energy is called a 'quantum' (plural 'quanta'), which is a Latin word meaning 'amount'. At relatively low temperatures, there is only enough thermal energy to excite low-frequency i.e. long wavelength quanta, whereas those of higher frequency are generated only when the temperature is higher. This is consistent with the general pattern of observation described above, but Planck's theory does even better than this. The formulae he developed on this basis actually produce a quantitative account of how much radiation is produced at each wavelength at a given temperature and these predictions agree precisely with the results of measurement.

The second set of phenomena that led to the quantum postulate is known as the 'photoelectric effect'. When light strikes a clean metal surface in a vacuum, electrons are emitted. These all carry a negative electric charge, so the stream of electrons constitutes an electric current. Applying a positive voltage to the metal plate can stop this current and the smallest voltage that is able to do so gives a measure of the energy carried by each electron. When such experiments are carried out, it is found that this electron energy is always the same for light of a given wavelength; if the light is made brighter, more electrons are emitted, but the energy carried by each individual electron is unaltered.

In 1905, Albert Einstein (at that time almost completely unknown to the scientific community) published three papers that were to have a revolutionary effect on the future of physics. One of these related to the phenomenon of 'Brownian motion', in which pollen grains in a liquid are seen to move at random when observed under a microscope: Einstein showed that this was due to them being bombarded by the atoms in the liquid and this insight is generally recognized to constitute the final proof of the existence of atoms. Another paper (the one for which he is most celebrated) set out the theory of relativity, including the famous relation between mass and energy. However, we are concerned with the third paper – for which he was awarded the Nobel Prize for physics – which offered an explanation of the photoelectric effect based on Planck's quantum hypothesis. Einstein realized that if the energy in a light wave is delivered in fixed quanta, then when light strikes a metal, one of these will transfer its energy to an electron. As a result, the energy carried by an electron will be equal to that delivered by a light quantum, minus a fixed amount required to remove the electron from the metal (known as the 'work function') and the shorter the wavelength of the light, the higher will be the energy of the emitted electron. When measurements of the properties of the photoelectric effect were analysed on this basis, it was found that they were in complete agreement with Einstein's hypothesis and the value of Planck's constant deduced from these measurements was the same as that obtained by Planck from his study of heat radiation.

An important additional observation was that, even if the intensity of the light is very weak, some electrons are emitted immediately the light is switched on, implying that the whole quantum is instantaneously transferred to an electron. This is just what would happen if light were composed of a stream of particles rather than a wave, so the quanta can be thought of as light particles, which are called 'photons'.

We therefore have evidence from the interference measurements that light is a wave, while the photoelectric effect indicates that it has the properties of a stream of particles. This is what is known as 'wave–particle duality'. Some readers may expect, or at least hope, that a book like this will explain to them how it is that light can be both a wave and a particle. However, such an explanation probably does not exist. The phenomena that exhibit these quantum properties are not part of our everyday experience (although it is a major aim of this book to show that their consequences are) and cannot be fully described using classical categories such as wave or particle, which our minds have evolved to use. In fact, light and other quantum objects are rarely completely wavelike nor fully particle-like, and the most appropriate model to use generally depends on the experimental context. When we perform an interference experiment with an intense beam of light, we generally do not observe the behaviour of the individual photons and to a very good approximation we can represent the light as a wave. On the other hand, when we detect a photon in the photoelectric effect we can usefully think of it as a particle. In both cases these descriptions are approximations and the light actually combines both aspects to a greater or lesser degree. Attempts to understand quantum objects more deeply have raised conceptual challenges and led to vigorous philosophical debate over the last hundred or so years. Such controversies are not central to this book, which aims to explore the consequences of quantum physics for our everyday experience, but we shall return to discuss them briefly in the last chapter, where we shall also discuss Schrödinger's famous, or notorious, cat.

matter waves

The fact that light, which is conventionally thought of as a wave, has particle properties led the French physicist Louis de Broglie to speculate that other objects we commonly think of as particles may have wave properties. Thus, a beam of electrons, which is most naturally imagined as a stream of very small bullet-like particles, would in some circumstances behave as if it were a wave. This radical idea was first directly confirmed in the 1920s by Davidson and Germer: they passed an electron beam through a crystal of graphite and observed an interference pattern that was similar in principle to that produced when light passes through a set of slits (cf. Figure 2.4). As we saw, this property is central to the evidence for light being a wave, so this experiment is direct confirmation that this model can also be applied to electrons. Later on, similar evidence was found for the wave properties of heavier particles, such as neutrons, and it is now believed that wave–particle duality is a universal property of all types of particle. Even everyday objects such as grains of sand, footballs or motorcars have wave properties, although in these cases the waves are completely unobservable in practice – partly because the relevant wavelength is much too small to be noticeable, but also because classical objects are composed of atoms, each of which has its own associated wave and all these waves are continually chopping and changing.

We saw above that in the case of light the vibration frequency of the wave is directly proportional to the energy of the quantum. In the case of matter waves, the frequency turns out to be hard to define and impossible to measure directly. Instead there is a connection between the wavelength of the wave and the momentum of the object, such that the higher is the particle momentum the shorter is the wavelength of the matter wave. This is discussed in more detail in Mathematical Box 2.3.

MATHEMATICAL BOX 2.3

As we saw in Chapter 1, momentum is defined as the mass (m) of a moving object multiplied by its velocity (v):

$$p = mv$$

De Broglie postulated that in the case of matter waves the wave–particle connection is 'wavelength equals Planck's constant divide by momentum':

$$l = h/p = h/(mv)$$

Planck's constant, h, is a fundamental constant of nature whose value equals 6.6×10^{-34} J s. Using this, we see that for an electron of mass around 10^{-30} kg moving at a (typical) speed of 10^6 m s^{-1}, its wavelength comes out at about 6×10^{-10} m, which is similar to that of typical X-rays. However, a grain of sand of mass about 10^{-8} kg moving at a speed of 1 mm s^{-1} has a wavelength of only 10^{-20} m, which renders its wave properties completely unobservable.

In classical waves, there is always something that is 'waving'. Thus in water waves the water surface moves up and down, in sound waves the air pressure oscillates and in electromagnetic waves the electric and magnetic fields vary. What is the equivalent quantity in the case of matter waves? The conventional answer to this question is that there is no physical quantity that corresponds to this. We can calculate the wave using the ideas and equations of quantum physics and we can use our results to predict the values of quantities that can be measured experimentally, but we cannot directly observe the wave itself, so we need not define it physically and should not attempt to do so. To emphasise this, we use the term 'wave function' rather than wave, which emphasises the point that it is a mathematical function rather than a physical object. Another important technical difference between wave functions and the classical waves we discussed earlier is that, whereas the classical wave oscillates at the frequency of the wave, in the matter-wave case the wave function remains constant in time.[3]

However, although not physical in itself, the wave function plays an essential role in the application of quantum physics to the understanding of real physical situations. Firstly, if the electron is confined within a given region, the wave function forms standing waves similar to those discussed earlier; as a result, the wavelength and therefore the particle's momentum takes on one of a set of discrete quantized values. Secondly, if we carry out experiments to detect the presence of the electron near a particular point, we are more likely to find it in regions where the wave function is large than in ones where it is small. This idea was placed on a more quantitative basis by Max Born, whose rule states that the probability of finding the particle near a particular point is proportional to the square of the magnitude of the wave function at that point.

Atoms contain electrons that are confined to a small region of space by the electric force attracting them to the nucleus. From what we said earlier, we could expect the associated wave functions to form a standing-wave pattern and we shall see shortly how this leads to an understanding of important properties of atoms. We begin this discussion by considering a simpler system in which we imagine an electron to be confined within a small box.

an electron in a box

In this example we consider the case of a particle, which we will assume to be an electron, trapped inside a box. By this we mean that if an electron is in the box, its potential energy has a constant value, which we can take to be zero. The electron is confined to the box because it is surrounded by a region of very high potential energy, which the electron cannot enter without breaching the principle of energy conservation. A classical analogy would be a ball inside a square box lying on the floor: provided the sides of the box are high enough, the ball cannot escape from the box, because to do so it would need to overcome gravity. We shall soon be considering the matter waves appropriate to this situation and we might compare these to the case of a pond or swimming pool, where the water is surrounded by a solid border: the solid shore is incapable of vibrating, so any waves generated must be confined to the water.

As a further simplification, we treat the problem as 'one-dimensional', by which we mean that the electron is confined to move along a particular direction in space so that motion in the other directions can be ignored. We can then make an analogy with waves on a string, which are essentially one-dimensional, because they can only move along the string. We now consider the form of the electron wave function. Because the electron cannot escape from the box the probability of finding it outside is zero. If we consider the very edge of the box, the probability of finding the particle at that point can have only one value, so the fact that it is zero outside the box means that it must also be zero just inside. This condition is very like that applying to a violin or guitar string and we saw earlier that this implies that the wave must be a standing wave with wavelength such that it fits into the space available (Figure 2.3). This is illustrated in Figure 2.5 and we see that the wavelength of the wave is restricted to one of the values corresponding to a whole number of half wavelengths fitting into the box. This means that only these particular values of the wavelength are allowed and, as the electron momentum is determined by the wavelength through the de Broglie relation, the momentum is also restricted to a particular set of values (see Mathematical Box 2.4). Remembering that the potential energy is zero and that the electron's kinetic energy depends only on its (known) mass and its momentum, we see that the total energy is similarly confined to one of a set of particular values – i.e. the energy is 'quantized' into a set of 'energy levels'. More detail is given in Mathematical Box 2.5, which contains expressions for the allowed energy values. These are also shown on Figure 2.5, where we see that the spacing between successive levels gets larger as the energy increases. We can now begin to understand some of the properties of atoms discussed towards the end of Chapter 1 on the basis of these results, but before doing so we shall use this example to discuss the idea of 'uncertainty' in quantum physics.

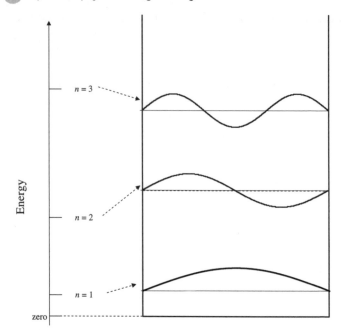

Figure 2.5 The energy levels and wave functions for the energy states of an electron in a box. Because the wave functions must be equal to zero at the edges of the box, the box length must equal a whole number of half wavelengths and this condition leads to the allowed values of the energy. The three states of largest wavelength and therefore lowest energy are shown. The numbers, n, are as in Mathematical Box 2.5.

Readers may well have come across reference to the 'Heisenberg uncertainty principle'. This is named after Werner Heisenberg, a pioneer of the ideas of quantum physics, who devised his own approach to the subject shortly before Schrödinger developed his equation. In general terms, the uncertainty principle states that it is impossible to know the exact values of two physical quantities, such as the position and momentum of a particle, at the same time. We can see how this works by referring to our example of the particle in the box. If we first consider its position, all we know is that the particle is somewhere in the box, and we define the uncertainty in position as the distance from the centre to the box edge, which is half the box size. Turning to momentum, if we consider a particle in the ground state, the wave function has the form of part of a wave whose wavelength is twice the box size: as the particle could be moving in either direction (left or right), the uncertainty in the momentum (defined similarly to that in position) is its maximum magnitude, which depends on the wavelength. It follows that if the box were larger, the uncertainty in position would be larger, but that in momentum would be smaller. If we multiply these quantities together, we find that the box size cancels out and the product equals Planck's constant (more details are given in Mathematical Box 2.4). The Heisenberg uncertainty principle states that the product of the uncertainties in position and momentum can never be smaller than a number approximately equal to one tenth of Planck's constant, and we see that this is indeed the case for our example. This is a general property of any wave function associated with a quantum state; we should note that the uncertainty principle is therefore a *consequence* of wave–particle duality and therefore quantum physics, rather than something additional to it.

Turning now to a comparison of the properties of our example with those of atoms, discussed in Chapter 1, we first note that the system has a lowest possible energy level, which is known as the 'ground state'. If therefore we had a number of identical boxes containing electrons, their ground states would also be identical. One of the properties of atoms that we could not explain classically was that all atoms of a given type have the same properties, and in particular that they all have the same lowest energy state. Through wave–particle duality, quantum physics has explained why such a state exists in the case of an electron in a box and we shall see shortly how the same principles apply to an electron in an atom.

Now consider what happens to the electron in a box when it changes from one allowed energy level to another – say from the first excited state to the ground state. To conserve energy, the energy lost must go somewhere and if we assume it is emitted in the form of a quantum of electromagnetic radiation, the wavelength of this radiation can be calculated from the difference between the energy levels using the Planck formula. We have all the information needed to calculate this in the case of an electron in a box whose length is about the diameter of an atom and this is

MATHEMATICAL BOX 2.4

We apply the results we found earlier for standing waves on a string (see Mathematical Box 2.2) to the case of the electron. This tells us that the wavelength of the wave functions associated with an electron in a box of length L, must have one of the values

$$l_n = 2L/n$$

where n is a whole number. It follows from de Broglie's postulate (Mathematical Box 2.3) that the magnitude of the electron momentum must have one of the values

$$p_n = h/l_n = nh/2L$$

We can use this to illustrate the Heisenberg uncertainty principle. When a physical quantity has a spread of possible values, we define its uncertainty as half the size of this spread. In the case of the position of a particle in a box, this quantity is d_x where

$$d_x = \frac{1}{2}L$$

And that in momentum[4] is

$$d_p = p_n = nh/2L \, d_{-p}$$

It follows that

$$d_x d_p = nh/4$$

The smallest value this can have is $h/4$, when $n = 1$. Heisenberg's uncertainty principle states that

$$d_x d_p \geq h/4\pi$$

where π is a mathematical constant whose value is about 3.142. Clearly our result is consistent with this.

done in Mathematical Box 2.5, where it is found that the radiation's wavelength is similar in size to that measured experimentally when a hydrogen atom makes a similar transition. Again we see how quantum physics accounts for atomic properties that we were unable to explain classically.

MATHEMATICAL BOX 2.5

Remembering that momentum equals mass times velocity and that the potential energy is zero in this case, the energy of the particle in a box is

$$E_n = \frac{1}{2}mv_n^2 = p_n^2/2m = (h^2/8mL^2)n^2$$

where we have used the expression for p_n derived in Mathematical Box 2.4.

If L is similar to the size of an atom (say 3×10^{-10} m), then, using the known value of the mass of an electron ($m = 10^{-30}$ kg),

$$E_n = 5 \times 10^{-19} n^2 \text{ J}$$

The change in energy when an electron moves from its $n = 2$ to its $n = 1$ state is

$$3h^2/8mL^2 = 1.1 \times 10^{-18} \text{ J}$$

If this energy is given to a photon, the frequency, f, of the associated electromagnetic wave will be this divided by h, and the corresponding wavelength is

$$l = c/f = 8mL^2 c/3h = 1.1 \times 10^{-7} \text{ m}$$

This is quite similar to the wavelength of the radiation emitted when a hydrogen atom makes a transition from its first excited state to its ground state, which is 1.4×10^{-7} m.

We should be encouraged that the numbers come out about the right size and we can at least tentatively believe that some properties of atoms result from the wave nature of their electrons. However, we should remember that there are still major differences between a real three-dimensional atom and our one-dimensional box. We saw in Chapter 1 that atoms consist of negatively charged electrons attracted to a positively charged nucleus so that the potential energy of attraction diminishes the further the electron is from the nucleus. The result is to confine the electron to the vicinity of the nucleus and we could expect the wave functions to be standing waves. However, not only is the atomic 'box' three-dimensional, but its shape is quite different from that discussed above, so we may not be fully convinced of the correctness of our approach before we have applied it to the actual atomic potential. We shall return to this in more detail shortly.

varying potential energy

So far we have considered the matter waves associated with particles propagating in free space or trapped in a one-dimensional box. In both these cases, the particle moves in a region where the potential energy is constant, so, if we remember that the total energy is conserved, the kinetic energy and hence the particle's momentum and speed must be the same wherever it goes. In contrast, a ball rolling up a hill, for example, gains potential energy, loses kinetic energy and slows down as it climbs. Now we know that the de Broglie relation connects the particle's speed to the wavelength of the wave, so if the speed stays constant, this quantity will also be the same everywhere, which is what we have implicitly assumed. However, if the speed is not constant, the wavelength must also vary and the wave will not have the relatively simple form we have considered so far. Hence, when a particle moves through a region where the potential energy varies, its speed and hence the wavelength of the wave function will also change.

In general, the analysis of a situation where the potential energy varies requires a study of the mathematical equation that controls the form of the wave in the general case. As mentioned earlier, this equation is known as the 'Schrödinger equation'. In the examples discussed above, where the potential is uniform, the solutions of the Schrödinger equation have the form of travelling or standing waves and our fairly simple approach is justified. A full understanding of more general situations is mathematically quite challenging and not appropriate to this book. Nevertheless, we can gain a lot of insight on the basis of our earlier discussion if we are prepared to take some of the details on trust. We will shortly apply this to a study of the structure of atoms and in the next chapter we shall see that simple travelling and standing waves can represent the motion of electrons in metals. First, however, we will try to deepen our understanding of the wave nature of a particle moving in a varying potential by considering two further examples.

quantum tunnelling

We first consider the case of a particle approaching a 'potential step'. By this we mean that the potential increases suddenly at a particular point, as is illustrated in Figure 2.6. We are particularly interested in the case where the energy of the approaching particle is smaller than the step height, so from a classical point of view we would expect the particle to bounce back as soon as it reaches the step, and then to move backwards at the same speed. Much the same thing happens when we apply quantum physics, but there are important differences, as we shall see. First we consider the form of the matter wave. On the basis of our earlier discussion, we expect particles approaching the step to be represented by travelling waves moving from left to right, whereas after they bounce back, the wave will be travelling from right to left. In general we do not know what the particle is doing at any particular time, so the wave function to the left of the step will be a combination of these, and this is confirmed when the Schrödinger equation is solved mathematically. What is of real interest is the form of the wave to the right of the step. Classically, there is no probability of finding the particle there, so we might expect the wave function to be zero in this region. However, when we solve the Schrödinger equation we find that, as shown in Figure 2.6(a), the calculated wave function does not become zero until some way to the right of the step. Remembering that the intensity of the wave function at any point represents the probability of finding a particle at that point, we see that quantum physics predicts that there is a finite chance of finding it in a region where it could never be if classical physics were the whole story.

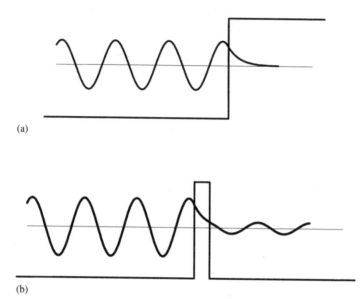

(a)

(b)

Figure 2.6 The strong straight lines in (a) represent a potential step. The wave function for a particle approaching the step is also shown; it penetrates the step, giving a probability of finding the particle in a region that is forbidden classically. The corresponding case for a narrow barrier is shown in (b): the wave function penetrates the barrier so that there is a probability of the particle emerging on the right-hand side, where it could never be classically. This is known as 'quantum mechanical tunnelling'.

It turns out to be impossible to test the above prediction directly, since placing any kind of detector inside the barrier would effectively change the form of the potential, but we can test it indirectly if we consider the slightly different situation illustrated in Figure 2.6(b). Instead of a step, we now have a 'barrier' where the potential steps back down to zero a little to the right of the upward step. When the Schrödinger equation is solved in this situation, we find that the form of the wave function to the left of the barrier and inside it is very similar to that just discussed in the case of the step. However, there is now a travelling wave of comparatively small, but finite, amplitude to the right of the barrier. Interpreting this physically, we conclude that there is a small probability that a particle approaching the barrier from the left will not bounce back but will emerge from the other side. This phenomenon is known as 'quantum mechanical tunnelling' because the particle appears to tunnel through a barrier that is impenetrable classically.

There is a wide range of physical phenomena that demonstrate quantum tunnelling in practice. For example, in many radioactive decays, where 'alpha particles' are emitted from the nuclei of some atoms, the probability of this happening for a particular atom can be very low – so low in fact that a particular nucleus will wait many millions of years on average before decaying. This is now understood on the basis that the alpha particle is trapped inside the nucleus by the equivalent of a potential barrier, similar in principle to that discussed above. A very low amplitude wave exists outside the barrier, which means that there is a very small (but non-zero) probability of the particle tunnelling out.

In recent years, quantum tunnelling has been dramatically exploited in the scanning tunnelling microscope. In this device, a sharp metal point is held just above a metal surface. As a result, electrons tunnel through the barrier separating the metal point from the surface and a current flows. Referring back to Figure 2.6, we see that the wave function at the right-hand side of the barrier gets rapidly smaller as the barrier thickness increases, which implies that the tunnelling current reduces sharply as the distance between the metal point and the plate increases. If the point is now scanned across an uneven metal surface, the variations in tunnelling current provide information about this unevenness and a map of the surface results. This technique has been developed to the point where the unevenness associated with individual atoms can be detected, and an example of this is shown in Figure 2.7. Scientists' ability to observe and manipulate individual atoms using scanning tunnelling microscopy and other similar techniques has opened up a whole new field of science and technology known as 'nanoscience'.

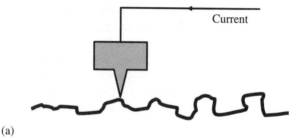

(a)

(b)

Figure 2.7 (a) A scanning tunnelling microscope moves a sharp point across a surface and detects the tunnelling current into the surface. This varies strongly with the distance of the point from the surface so any unevenness can be detected. The picture (b) shows an image of part of the surface of a crystal of silicon; the bright peaks correspond to individual atoms. Photograph supplied by P.A. Sloan and R.E. Palmer of the Nanoscale Physics Research Laboratory in the University of Birmingham, UK.

a quantum oscillator

The second example we consider is a particle moving in a parabolic potential, as illustrated in Figure 2.8. In the classical case, the particle would oscillate regularly from one side of the potential

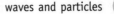

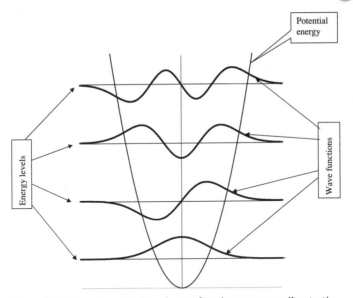

Figure 2.8 The energy levels and wave functions corresponding to the four lowest energy states of a particle moving in a parabolic potential. The wave functions have been drawn so that their zeros are at the corresponding energy levels. Note that the 'effective box size' is larger the higher the state and that the wave function penetrates the classically forbidden area, in a way similar to that in the potential step illustrated in Figure 2.6(a).

well to the other with a frequency determined by its mass and the shape of the well. The size, or 'amplitude', of the oscillation is determined by the particle's energy: at the foot of the well all this energy is kinetic, while the particle comes to rest at the limits of its motion, where all the energy is potential. The wave functions are obtained by solving the Schrödinger equation, and it is found that, just as in the case of a particle in a box (Figure 2.5), stand-ing-wave solutions are possible only for particular values of the energy. The energy levels and the wave functions associated with them are illustrated in Figure 2.8. There are important similarities and important differences between these and the corresponding standing waves shown in Figure 2.5. First the similarities: in both cases, the wave function corresponding to the lowest energy state is represented by a single hump that reaches a maximum in the centre; the next highest state has two humps, one positive and the other negative with the wave function crossing the axis and so on. Now the differences. First, the width occupied by the wave is the same for all states in the case of the box, but varies in the oscillator case, because as the total energy increases, so does the width of the region in which the total energy is positive: roughly speaking, we can say that the effective width of the box is different for the different energy levels. Secondly, the wave does not go to zero immediately the limits of the classical motion are reached, but penetrates the 'classically forbidden' region to some extent in a manner similar to the case of a particle approaching a step (cf. Figure 2.6[a]). This is discussed in more detail in Mathematical Box 2.6.

By studying this example, the reader will hopefully have appreciated how many of the features of such a problem can be deduced from an understanding of matter waves in a constant potential, although the details require a more mathematical approach. We shall now try to apply these principles to understand some of the quantum physics of real atoms.

the hydrogen atom

The simplest atom is that of the element hydrogen, which consists of a single negatively charged electron bound to a positively charged nucleus by the electrostatic (or 'Coulomb') force, which is strong when the electron is close to the nucleus and steadily reduces in strength when the electron is further away. As a result, the potential energy is large and negative near the nucleus and gets closer to zero as we move away from it (cf. Figure 2.9). The examples discussed so far have all been one-dimensional, meaning that we have implicitly assumed that the particle is constrained to move along a particular direction (from left to right or vice versa in our diagrams). However, atoms are three-dimensional objects and we will have to take this into account before we can understand them fully. An important simplifying feature of the

MATHEMATICAL BOX 2.6

We saw earlier that in the case of a particle in a box the energy levels increased rapidly as we went up the ladder, having the values

$$E_n = h^2 n^2 / 8mL^2$$

where L is the size of the box. As is clear from this formula and illustrated in Figure 2.5, the spacing between successive levels is larger the higher these levels are.

In the oscillator case, the potential energy has the form

$$V = kx^2$$

where k is a constant. The width occupied by the wave therefore gets larger as the energy increases, as illustrated in Figure 2.8. We can make an approximate comparison with the particle in the box if we assume that the effective size of the box is the width of the wave function and is therefore larger for higher energy levels. Putting all this together, we can predict that the spacings between the oscillator energy levels will increase less rapidly at high energies than do those of the box. This prediction is confirmed when the energy levels of the oscillator are calculated by solving the Schrödinger equation, which produces a set of evenly spaced energy levels in the case of the oscillator. The actual expression is

$$E_n = \left(n + \frac{1}{2} \right) hf$$

where f is the classical frequency of oscillation and n is a positive integer or zero.

hydrogen atom is that the Coulomb potential is 'spherically symmetric' – i.e. it depends only on the distance between the electron and the nucleus – whatever the direction of this separation. A consequence is that many of the wave functions associated with the allowed energy levels have the same symmetry; we shall discuss these first and return to the others later.

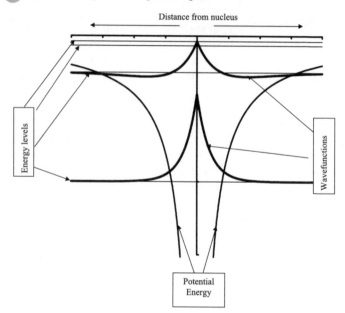

Figure 2.9 The hydrogen atom. The diagram shows the potential energy along with the four lowest energy levels and the wave functions corresponding to the lowest two of these; the wave functions have been drawn so that their zeros are at the corresponding energy levels. Note that the zero of energy corresponds to the top line in the diagram.

The Coulomb potential confines the electron to the vicinity of the nucleus in much the same way as the square box and the oscillator potential confine the particle in the examples discussed above. We saw how the effective box width in the oscillator case was larger for states of higher energy, which in turn meant that the energies of the higher states did not increase as fast as in the case of the square box. If we compare the shape of the Coulomb potential in Figure 2.9 with that of the oscillator in Figure 2.8, we see that the potential width increases even more rapidly with energy in the Coulomb case. Applying the same reasoning as in the oscillator case, we should expect the energy levels to increase even more slowly as we go up the ladder. This is indeed what happens and the energy levels come out to be $-R$, $-R/4$, $-R/9$, $-R/16$ ... where R is a constant known as the 'Rhydberg' constant, after the Swedish scientist who worked on atomic spectra towards the end of the nineteenth century. Notice that these numbers are negative because we measure the energy from a zero level that corresponds to the electron and nucleus being very far apart.

When an atom moves from one energy level to another, the energy is absorbed or emitted as a photon of radiation, whose frequency is related to the energy change by the Planck relation. The pattern of frequencies calculated in this way from the above pattern of energy levels is the same as that observed experimentally when electrical discharges are passed through hydrogen gas. A full solution of the Schrödinger equation predicts a value for the constant R in terms of the electron charge and mass and Planck's constant, and this value agrees precisely with that deduced from the experimental measurements; this is discussed in more detail in Mathematical Box 2.7. We now therefore have complete quantitative agreement between the predictions of quantum physics and the experimental measurements of the energy levels of the hydrogen atom.

We have used the principle of wave–particle duality to obtain the quantized energy levels, but how are we to interpret the wave function that is associated with each level? The answer to this question lies in the Born rule stated earlier: the square of the wave function at any point represents the probability of finding the electron near that point. A model of the atom consistent with this is that, in this context, the electron should be thought of not as a point particle but as a continuous distribution spread over the volume of the atom. We can envisage the atom as a positively charged nucleus surrounded by a cloud of negative charge whose concentration at any point is proportional to the square of the wave function at that point. This model works well in many situations, but should not be taken too literally: if we actually look for the electron in the atom, we will always find it as a point particle. On the other hand, it is equally wrong to think of the electron as being a point particle when we are not observing its position. In quantum physics, we use models, but do not interpret them too literally. We return to this in our discussion of the conceptual principles of the subject in Chapter 8.

So far we have discussed only states where the wave has spherical symmetry: i.e. it has the same value at the same distance from the nucleus, whatever the direction. However, there are other

MATHEMATICAL BOX 2.7

The hydrogen atom energy levels can be expressed by the general formula

$$E_n = -R/n^2$$

where n is an integer and R is a constant. An expression for R in terms of Planck's constant (h), the mass (m) and charge (e) of the electron and the scaling constant for the Coulomb potential (k; cf. Mathematical Box 1.3) can be obtained by solving the Schrödinger equation. The result is

$$R = 2\pi^2 k^2 \, me^4 / h^2$$

Using the known values of the quantities on the right-hand side of this equation ($k = 9.0 \times 10^9 \, \mathrm{J\,m\,C^{-2}}$; $m = 9.1 \times 10^{-31} \, \mathrm{kg}$; $e = 1.6 \times 10^{-19} \, \mathrm{C}$; $h = 6.6 \times 10^{-34} \, \mathrm{J\,s}$), we find that $R = 2.2 \times 10^{-18} \, \mathrm{J}$.

The pattern of energy levels is as illustrated in Figure 2.9, which also shows the form of the wave function for the lowest three states. We should not be surprised to see that it oscillates, the number of humps increasing as the energy increases, though the shapes and sizes of the humps are much more varied than in the case of an electron in a box.

We are now in a position to compare these predicted energy levels with experiment. Remembering that the energy of the photons absorbed or emitted when an atom undergoes a transition from one state to another is just the difference between these energy states, and using the Planck relation, we get

$$f_{m,n} = (R/m^2 - R/n^2)$$

where $f_{m,n}$ is the frequency of the photon associated with a transition between the states of energy E_n and E_m. Thus, for example,

$$f_{1,2} = 3R/4h; \; f_{1,3} = 8R/9h; \; f_{2,3} = 5R/36h$$

The associated wavelengths are calculated from the frequencies in the usual way, giving (where c is the speed of light, $= 3.0 \times 10^8 \, \mathrm{m\,s^{-1}}$)

$$l_{1,2} = 4ch/3R; \; l_{1,3} = 9ch/8R; \; l_{2,3} = 36ch/5R$$

Again, putting the numbers in, we get

$$l_{1,2} = 1.2 \times 10^{-7}\,m; \; l_{1,3} = 1.0 \times 10^{-7}\,m; \; l_{2,3} = 6.5 \times 10^{-7}\,m$$

These agree with the values measured experimentally from observations of the light absorbed and emitted from hydrogen atoms. Moreover, this is still true when the quantities are calculated using the most precise values available for the physical constants, which typically means eight or nine decimal places.

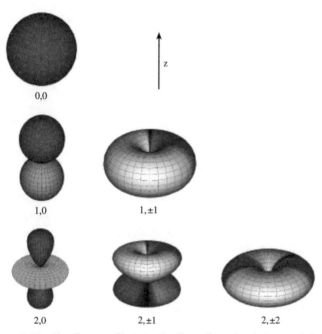

Figure 2.10 The diagrams illustrate the three-dimensional shapes of the wave functions corresponding to some of the hydrogen atom energy states. In each case, the nucleus is at the centre of the pattern, which represents the three-dimensional wave function.

states that do not have this simple property, but vary in direction. Some of these are illustrated in Figure 2.10, where we see that their shapes can be quite complex. The physical significance of these non-spherical states is that they correspond to the electron moving round the atom with some 'angular velocity' and associated 'angular momentum'. In contrast, the spherical waves correspond to the electron being spread around the volume of the nucleus but having no orbital motion. Given that there are all these non-spherical states, how come the spectrum of energy levels is not much more complex than we have discussed? By a lucky coincidence, it turns out that the energy of each of the non-spherical states is equal to that of one of the spherical states, so the simple picture we discussed earlier holds. If it had not been for this happy accident, the experimental spectrum of hydrogen would not have fitted the comparatively simple formula discussed above and the road to a successful quantum theory would have been considerably harder.

other atoms

Atoms other than hydrogen contain more than one electron, which causes further complications. Before we can address these, we have to consider another quantum principle, known as the 'Pauli exclusion principle' after its inventor Wolfgang Pauli. This states that any particular quantum state can contain more than one particle of a given type, such as an electron. This principle, although easily stated, can be proved only by using very advanced quantum analysis and we shall certainly not try to do this here. Before we can correctly apply the exclusion principle, however, we have to know about a further property possessed by quantum particles and known as 'spin'.

We know that the Earth spins on its axis as it moves in orbit around the sun, so if the atom were a classical object, we might well expect the electron to be spinning in a similar manner. This analogy holds to some extent, but once again there are important differences between the classical and quantum situations. There are two quantum rules that govern the properties of spin: first, for any given type of particle (electron, proton, neutron) the rate of spin is always the same; and, second, the direction of spin is either clockwise or anticlockwise about some axis.[5] This means that an electron in an atom can have one of only two spin states. Thus any quantum state described by a standing wave can contain two electrons provided they spin in opposite directions.

As an example of the application of the Pauli exclusion principle, consider what happens if we place a number of electrons in the box discussed earlier and illustrated in Figure 2.5. To form the state with lowest total energy, all the electrons must occupy the lowest possible energy levels. Thus, if we think of adding them to the box one at a time, the first goes into the ground state, as does the second with spin opposite to that of the first. This level is now full, so the third electron must go into the next highest energy level along with the fourth and the spins of these two electrons must also be opposite. We can continue adding up to two electrons to each energy state until all have been accommodated.

We now apply this process to atoms, first considering helium, which has two electrons. If we initially ignore the fact that the electrons exert a repulsive electrostatic force on each other, we can calculate the quantum states in the same way as we did for hydrogen, but allowing for the fact that the nuclear charge is doubled. This doubling means that all the energy levels are considerably reduced (i.e. made more negative), but otherwise the set of standing waves is quite similar to those in hydrogen and it turns out that this pattern is not greatly altered when the interactions between the electrons are included. The lowest energy state therefore has both electrons with opposite spin in the lowest. In the case of lithium with three electrons, two of these will be in the lowest state, while the third must be in the next higher energy state. The latter state can actually contain a total of six electrons: two of these occupy a state of spherical symmetry while the others fill three separate non-spherical states. A set of states with the same value of n is known as a 'shell' and if electrons occupy all these states, it is called a 'closed shell'. Thus lithium has one electron outside a closed shell, as does sodium with eleven electrons – i.e. two in the $n = 1$ closed shell, eight in the $n = 2$ closed shell and one electron in the $n = 3$ shell. It is known that many of the properties of sodium are similar to those of lithium and similar correspondences between the properties of other elements underlie what is known as the 'periodic table' of the elements.

The whole structure of the periodic table can be understood in terms of the atomic shell structure, which in turn is a consequence of the properties of the quantum waves associated with the electrons.

Although the above considerations allow us to describe the electronic structure of atoms in some detail, it is much harder to make a precise calculation of the energy levels. This is because if the atom contains more than one electron, there is a repulsive interaction between them (because the signs of their charges are the same) as well as an attractive force towards the nucleus. Even in the case of helium, with two electrons, the Schrödinger equation cannot be solved to give algebraic expressions for the allowed energies and the wave functions, and the problem becomes much harder as the numbers of electrons increase. Moreover, the exact agreement between the energies of spherical and non-spherical states only holds in the case of hydrogen, so the spectra of the other atoms are generally much more complex. However, modern computational techniques have largely taken over where traditional mathematics has failed. When applied to any atom, these produce values for the allowed energy levels and numerical representations of the wave functions which are in good agreement with experiment. All the evidence points to quantum physics providing a complete description of the physical properties of matter at the atomic scale.

summary

This chapter has introduced the main ideas of quantum physics, which will be developed and applied to various physical situations in the chapters to come. Readers would be well advised to ensure that they understand these basic principles, which are summarised below.

- Examples of classical waves are water waves, sound waves and light waves. They are all typified by a *frequency*, which determines how many times per second any point on the wave vibrates, and a *wavelength*, which measures the repeat distance along the wave at any time.

- Waves have the form of *travelling waves* or *standing waves*.
- Travelling waves move at a speed determined by the frequency and the wavelength.
- Because standing waves result from a wave being confined to a region in space, the wavelength and hence the frequency of a standing wave is restricted to have one of a set of allowed values. This is exemplified in the notes produced by musical instruments.
- Although there is evidence that light is a wave, in some circumstances it behaves as if it were a stream of particles, known as light quanta or 'photons'.
- Similarly, quantum particles such as electrons behave in some contexts as if they were waves.
- When an electron is confined by a potential, such as a 'box', the matter waves are standing waves with particular wavelengths, which in turn cause the electron energy to be quantized – i.e. to have one of a set of particular values.
- When a quantum system moves from one energy level to another, the change in energy is provided by an incoming photon or given to an outgoing photon.
- The wave properties of quantum particles enable them to tunnel through potential barriers that they could not surmount classically.
- The calculated and measured energy levels of the hydrogen atom agree precisely, which is strong evidence for the correctness of quantum physics
- The Pauli principle states that no two electrons can occupy the same quantum state. Because an electron can be in one of two spin states, this means that each standing wave can contain up to two electrons.

notes

1. A condition of this kind is known as a 'boundary condition'.
2. In the case of a stretched string, the wave speed is related to the tension on the string and its mass. Both are adjusted in many instruments: e.g. a violinist alters the tension when tuning their instrument, and heavier and lighter strings are used to produce lower and higher notes, respectively.

3. Technically, it is the *magnitude* of the wave function that stays constant, while its *phase* oscillates. However, this oscillating phase plays little, if any, part in determining the properties we shall be discussing. The magnitude of the wave function can also vary, but only in circumstances where the energy of the particle is not well defined, and I shall not be discussing these in this book.

4. Because the wave is cut off at the edges of the box, it turns out that the magnitude of the momentum has a spread of values rather than just that given by p_n. However, the size of this spread is similar to d_p as defined above.

5. The idea that an electron is literally spinning should be thought of as a semi-classical model rather than a literal description. What we call 'spin' is a property that emerges from an advanced mathematical treatment combining the principles of quantum physics and relativity. The basic result that an electron can be in one of two spin states is also a consequence of this.

power from the quantum

In this chapter we shall discover how quantum ideas play an important role in the physics of the generation of energy. As soon as humankind discovered fire and how to use it, quantum physics was directly involved in energy production and this is still true for many of the forms of energy generation that play an essential role in modern life. We burn petrol in our cars and use gas or oil to heat our homes. Much of the power we use reaches our homes in the form of electricity, although it is important to remember that this is not in itself a source of power, but only a method of transferring energy from one place to another. Electricity is generated at a power station, from the energy stored in its fuel, which may be a 'fossil fuel' such as coal, oil or gas; a nuclear fuel such as uranium or plutonium; or a source of 'sustainable' energy, such as solar, wind or wave power. Of all these, only wind and wave power do not depend directly on quantum physics.

chemical fuels

A fuel such as wood, paper, oil or gas contains many hydrocarbons, which are compounds consisting mainly of hydrogen and carbon atoms. When these are heated in air, the hydrogen and carbon combine with oxygen from the air to make water and carbon dioxide, respectively. In the process, energy is released in the form of heat, which can then be used to produce electricity in a power station or to power a motor vehicle, for example.

61

To see how this depends on quantum physics, we start with the simplest example of chemical combination, which is two hydrogen atoms coming together to form a hydrogen molecule – see Figure 3.1. The hydrogen atom contains one electron attracted to a

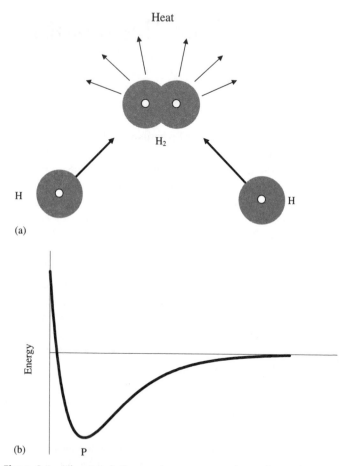

Heat

H_2

H

H

(a)

Energy

(b)   P

Figure 3.1 When two hydrogen atoms come together to form a hydrogen molecule (a), the total energy of the system is reduced and the extra energy is released as heat. The graph (b) shows the variation of the energy of the system as the separation between the hydrogen atoms changes. The final state of the molecule corresponds to the point of lowest energy marked P.

proton whose charge is equal and opposite to that of the electron, so a hydrogen molecule is composed of two protons and two electrons. In Chapter 2 we showed how the wave properties of the electron result in the energy of the hydrogen atom being quantized so that its value equals one of a set of specific energy levels; in the absence of excitation, the atom is in the lowest of these energy states, known as the 'ground state'. Now consider how the total energy of the system will be affected if we bring two hydrogen atoms towards each other. We first consider the potential energy which changes in three ways. First, it increases because of the electrostatic repulsion between the two positively charged protons; secondly, it decreases because each electron is now subject to attraction by both protons; thirdly it increases because of the repulsion between the two negatively charged electrons. In addition, the kinetic energy of the electrons decreases because the electrons are able to move around and between the two nuclei, so the size of the effective box confining them is increased. (We saw in Chapter 2, when we discussed the quantum behaviour of a particle in a box, that the larger the box, the lower is the kinetic energy of the ground state.) We also note that the Pauli exclusion principle allows both electrons to occupy the ground state, provided they have opposite spin. The net effect of all these changes depends on how far the atoms are apart: when they are widely separated, there is little change in the total energy, and when they are very close, the electrostatic repulsion between the nuclei dominates. However, at intermediate distances, there is an overall reduction in the total energy and this reduction is at its greatest when the protons are about 7.4×10^{-10} m apart (cf. Figure 3.1[b]). At this point, the difference between the energy of the molecule and that of the widely separated hydrogen atoms equals about one third of the ground-state energy of the hydrogen atom. Where does this surplus energy end up? The answer is that some of it goes into the kinetic energy of the moving molecule, while the rest is given off in the form of photons. Both are effectively forms of heat, so the overall effect is a rise in temperature, which is just what we expect from a fuel.

The above example illustrates the principle of how energy can be released by bringing atoms together to form molecules, but the particular case of hydrogen is not in practice a useful source of energy, because any hydrogen gas we have on Earth is already composed of molecules. A more practical example is the combination of hydrogen and oxygen to make water: the ground state energy of the water molecule is less than the total ground state energies of the single oxygen and two hydrogen atoms that are its constituents. However, like hydrogen, oxygen gas is also composed of diatomic molecules and if we simply mix hydrogen and oxygen together at room temperature, nothing happens. This is because, before they can combine to form water, the oxygen and hydrogen molecules must first be split into their constituent atoms, which requires an input of energy from an external source. However, once a few water molecules have formed, the energy released in this process is more than sufficient to split apart some more hydrogen and oxygen molecules, and the process very quickly becomes self-sustaining. An example of this is lighting a gas flame in a laboratory or kitchen, using a match: the high temperature produced by the match splits some of the nearby hydrogen and oxygen molecules and the resulting atoms combine to form water molecules, with a release of energy that heats more of the gas to the point where it too can ignite. The process is then self-sustaining and the heat produced can be used to warm a house, boil a kettle, etc.

The principles involved in this example underlie all useful chemical fuels and indeed nuclear energy, as we shall see later. A hydrocarbon fuel, such as oil or gas, contains molecules composed primarily of carbon and hydrogen, which have remained stable for a long time – perhaps millions of years. This stability is maintained even when the compounds are exposed to air at room temperature, but once energy is supplied to split the molecules, the atoms rearrange themselves into a mixture of water and carbon dioxide with the release of energy. The principles involved are those of quantum physics: the total energy of the quantum ground states of the water and carbon dioxide molecules is less than that of the initial hydrocarbon molecules. However, to initiate this change, energy must be supplied; once the mixture has been heated to a sufficiently high temperature, the process becomes self-sustaining and (unless the process is extinguished) energy continues to be released until the fuel is exhausted.

nuclear fuels

The principles of nuclear power are remarkably similar to those underlying the burning of chemical fuels, although the amounts of energy involved in the nuclear processes are very much greater. As we saw in Chapter 1, the nucleus of an atom is made up of a number of protons and neutrons bound together by the strong nuclear force. The structure of the nucleus is also subject to the rules of quantum physics, although the details are rather more complex than the atomic case discussed in Chapter 2. This is because the latter is dominated by the attraction of the electrons to the heavier nucleus, whereas the interactions between the protons and neutron inside the nucleus are all of similar mass. However, the outcomes are quite similar in both cases: like the atom, the energy of the nucleus is quantized into a set of energy levels, the lowest of which is known as the 'ground state'.

Closely analogous to the earlier example of the combination of two hydrogen atoms into a hydrogen molecule is the 'fusion' of two hydrogen nuclei into a single nucleus. The hydrogen nuclei are protons and the resulting nucleus is known as 'deuterium'. As discussed in Chapter 1, deuterium is an isotope of hydrogen whose nucleus is composed of a proton and a neutron and which makes up about 0.02% natural hydrogen gas. As the neutron carries no charge, the extra positive charge must go somewhere and it is actually carried off by the emission of a positron (which is the same as an electron but with a positive charge) and a neutrino (a very small neutral particle). The ground state energy of the deuterium nucleus is considerably lower than that of two protons, so we might have expected that all the protons in the universe would have been fused into deuterium nuclei many years ago, in the same way that practically all hydrogen atoms have formed hydrogen molecules. The fact that this has not happened is due to the electrostatic repulsion between the two positively charged protons. The strong nuclear force that binds the protons together in the nucleus is a short-range force that is appreciable only when the nucleons are within about 10^{-15} m of each other. As the protons come together, the electrostatic repulsion increases to a very large value before the nuclear force kicks in, creating a potential barrier as illustrated in Figure 3.2. Classically, this barrier would completely prevent the protons ever combining, but, in principle, they can penetrate it by quantum tunnelling (see Chapter 2). Detailed calculations show that the probability of this happening is very low unless the protons are moving towards each other at very high speed, in which case the effective tunnelling barrier is both lower and narrower – see Figure 3.2. We therefore have something akin to the ignition process discussed in the last section: to get the fusion energy out we must first put energy in, but the amount of energy required in this case is equivalent to raising the temperature by several million degrees. The energy obtained as a result of the fusion process is also high: that released from the fusion of two protons is about ten million times that associated with the formation of a hydrogen molecule from two hydrogen atoms.

One place where temperatures as high as a million degrees occur naturally is the sun, and indeed nuclear fusion is the process that keeps the sun shining. Many other fusion processes besides that of two protons to form deuterium take place there and the end point is the most stable nucleus of all – that of iron. Fusion is also one of the principles underlying nuclear weapons such as the 'hydrogen bomb'. In this case, the ignition is achieved from a nuclear explosion generated using atomic fission, which will be discussed shortly. This heats the material to a temperature high enough for fusion to start, after which it is self-sustaining and an enormous explosion results. The generation of controlled fusion power that could be used for peaceful purposes has been an aim of nuclear researchers for over fifty years.[1] The technological challenges are enormous and the machines needed to generate and maintain the required temperatures are huge and amount to an investment of many billions of pounds. International collaborations such as the JET[2] project have been formed to pursue this and it is now believed that a machine capable of producing significant amounts of fusion power will be built during the first half of the twenty-first century. Towards the end of the twentieth century, some excitement was generated by reports of 'cold fusion', meaning fusion energy liberated without first supplying heat. This work has been largely discredited, but some efforts continue in this direction.

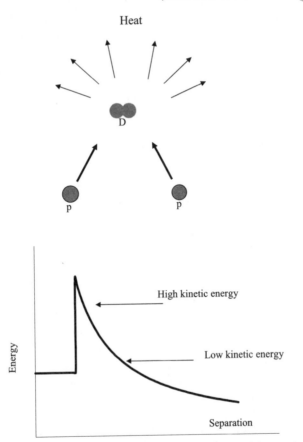

Figure 3.2 When two protons fuse together to form a deuterium nucleus (De), energy is released and a positron and neutrino (neither are shown above) are emitted. However, before this can occur, they must tunnel through the potential barrier created by the electrostatic repulsion. The probability of this happening is very low, unless the protons approach each other at high speed and therefore high kinetic energy, which makes the effective barrier lower and narrower. Note: the scale of this diagram is about two thousand times larger than that in Figure 3.1 and the energy released is about a million times greater.

Another form of nuclear power that has been well established for many years is known as 'fission', which means the splitting apart of a nucleus into smaller fragments. We have seen that the energy of the ground state of the deuterium nucleus is lower than that of two protons and this trend continues as we form heavier atoms, until we come to iron whose nucleus contains twenty-six protons and thirty neutrons. Beyond this, the trend is reversed and splitting a nucleus of a heavier element into pieces may well result in a lowering of the total ground-state energy and the release of energy. To understand this in a little more detail, consider an example where a heavy nucleus splits into two equal-sized fragments (Figure 3.3). Before this occurs, the two parts are held together by the strong nuclear force, but once they are separated the electrostatic repulsion between the positively charged fragments takes over, pushing them further apart into a yet lower energy state and releasing the surplus energy. We can think of this as the reverse of the process of fusing two deuterons to make an alpha particle, except that in the present case the energy of the widely separated state is less than that of the ground state of the united nucleus so energy is released when the nucleus splits – i.e. undergoes fission. However, an initial energy barrier has to be surmounted before fission can occur, which would seem to imply a need to inject energy into the system. To do so would be even more impractical than in the case of fusion, so a different approach is needed.

The key to the initiation of fission lies in some of the detailed properties of nuclear structure. When this is analysed, taking full account of the nuclear forces and the electrostatic repulsion, it is found that stable bound states occur only for a limited number of particular combinations of protons and neutrons. An example of an unstable nucleus is U^{236}, which is a uranium nucleus containing 92 protons and 144 neutrons; if this were ever created, it would immediately fly apart because there is no energy barrier preventing this 'spontaneous fission'. In contrast, a nucleus formed with one fewer nucleon (i.e. U^{235}) is relatively stable and a small amount (a little less than one per cent) occurs in natural uranium. To induce fission in U^{235}, therefore, what we can do is to add another neutron to the nucleus. As a neutron carries no charge, there is no energy barrier preventing it from entering a U^{235}

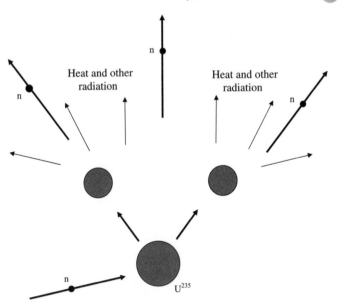

Figure 3.3 When a neutron enters a nucleus of the uranium isotope U^{235}, it becomes unstable and undergoes fission into two fragments along with some extra neutrons and other forms of radiation, including heat. The released neutrons can cause fission in other U^{235} nuclei, which can produce a chain reaction.

nucleus, so converting it to U^{236}, which then undergoes fission. We note that the neutron does not have to possess any extra energy for this to happen; indeed, if it is moving too fast, it is likely to pass by the U^{235} nucleus without interacting with it. We do not have to ignite a fission process with energy, but to start it off we do have to supply neutrons.

I have described fission as the splitting of a heavy nucleus into two smaller fragments, but in practice the process is appreciably more complex. In particular, radiation in the form of high-energy alpha particles (i.e. He^4 nuclei) is produced as well as some free neutrons. These are then available to induce fission in other U^{235} nuclei that may be nearby. This process can then multiply and we get a 'chain reaction' in which all the nuclei in a piece of U^{235} undergo fusion in a very short time, resulting in the huge explosion associated with an 'atomic' bomb. To start it off, we seem to need to seed it with some neutrons, but in fact there is a small probability of U^{235} undergoing spontaneous fission and producing a few neutrons. Some of these may strike other uranium nuclei, inducing them to split also. If the piece of uranium is small in size, many of the neutrons will escape from it and the process will not be self-sustaining, but in the case of a large sample the process will multiply and form a chain reaction. Thus, all that is needed for a nuclear explosion is to create a 'critical mass' of U^{235} – i.e. to bring a sufficient quantity together and hold it in place until the chain reaction is complete. In an atomic bomb, this is achieved by bringing two or more smaller masses together very quickly using a conventional explosive; in contrast, a nuclear reactor is designed so that the fission process is controlled in order that the energy released can be used to generate electricity. Neither process can be realized without materials that contain a sufficiently high concentration of U^{235}. As we saw above, natural uranium contains only about one per cent of U^{235} and the material has to be 'enriched' to increase this to around twenty per cent before it can be used in a reactor, while 'weapon-grade' uranium contains up to ninety per cent U^{235}. No doubt fortunately, this enrichment process is difficult and expensive and constitutes one of the major technological barriers to the use of nuclear energy, particularly in the weapons field.

There are several different designs of nuclear reactor. Figure 3.4 illustrates the principles of one of these – a 'pressurized water' reactor. Rods of enriched uranium are held in a vessel along with water under high pressure. This acts as a 'moderator', meaning that the kinetic energy of the neutrons is reduced when they collide with the water molecules, which rebound taking some of the neutron energy with them, and raising the temperature of the water in the process. It turns out that slow neutrons are considerably more efficient at inducing fission in uranium nuclei than are neutrons of higher energy, so the moderation leads to an increase in the efficiency of the fission process. Because the water in the reactor is at very high pressure, it can be heated to a high temperature without boiling. The high-temperature water is pumped out of the pressure vessel and its energy is used to heat water at normal pressure to produce steam, which drives a turbine to

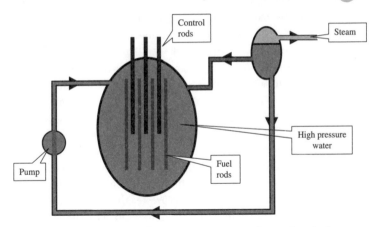

Figure 3.4 In a high-pressure water reactor, energy is transferred when the neutrons created by the fission process collide with the water molecules. As a result, the water is heated up and the slowed-down neutrons cause further fission. Rods made of neutron-absorbing material are raised or lowered to control the rate of the reaction. The high-pressure hot water is used to create steam, which in turn generates electricity.

generate electricity, while the cooled high-pressure water is recirculated. Rods of a material such as boron that absorbs neutrons passing into it can be lowered into the water; this reduces the number of neutrons available to induce further fission and so allows the rate of the reaction to be controlled.

Both fission and fusion produce energy by inducing transitions from quantum states of high energy to other more stable lower energy states. The laws of quantum physics and the wave properties of the neutrons and protons in turn determine the energies of these states. Once again the practical importance of quantum physics is demonstrated.

Some readers may have heard that Einstein's discovery of the equivalence of mass and energy is essential to nuclear energy and will be surprised that it has not been mentioned so far. The reason for this is a very common misunderstanding. Returning to the fusion example, it is true that the mass of the deuterium nucleus is a little less than the total mass of two protons and that the energy produced can be calculated from this missing mass using Einstein's famous equation, $E = mc^2$. However, the mass loss is not the *cause* of the energy change (which is the strong nuclear force and the electrostatic repulsion) but an inevitable *consequence* of it. Einstein's equation means that changes in mass accompany all energy changes, including, for example, those associated with the chemical reactions we discussed earlier. Thus, when we burn hydrogen and oxygen to form water, the mass of a resulting water molecule is a little less than the total masses of the hydrogen and oxygen atoms it is composed of. However, in the combustion case these changes are extremely small and difficult to measure (typically, less than one part in a hundred million) whereas in the case of nuclear energy they are much more significant: e.g. the total mass of two protons is about 0.05% greater than that of the products created when they fuse to form deuterium. This difference is relatively easy to measure and, historically, it was probably this fact, interpreted in the light of Einstein's relation between mass and energy, that led scientists to realize how much energy is associated with the strong nuclear force. However, this does not change the fact that it is the energy change that causes the mass change rather than the other way around.

green power

During the last twenty or so years of the twentieth century and since, we have become more and more conscious of the fact that our exploitation of the Earth's energy sources has given rise to considerable problems associated with pollution and the like. Some initial concerns were focused on nuclear energy, where the inevitable radiation accompanying nuclear processes and the disposal of radioactive waste products constitute hazards, which some feared could not be controlled. This was exacerbated by a small number of quite major nuclear accidents, particularly that in Chernobyl in the Ukraine, which released a considerable amount of radioactive material across Europe and beyond.

More recently, however, the long-term consequences of more traditional methods of energy production have become clear. Chief among these is the possibility of climate change associated with 'global warming': there are strong indications that the burning of fossil fuels is resulting in a gradual rise in the Earth's temperature, which in turn could result in the melting of the polar ice caps, a consequent rise in sea levels and the flooding of significant parts of the Earth's inhabited areas. There is even the possibility of a runaway process in which heating would result in more heating until the Earth became completely uninhabitable. In the face of such predicted disasters, there has been a rapid rise in interest in alternative 'sustainable' forms of energy production. In this section, we shall first discuss how quantum physics plays a role in causing the problem of global warming through the 'greenhouse effect' and how it can also contribute to some of the sustainable alternatives.

The greenhouse effect is so named because it mimics the processes that control the behaviour of a glass greenhouse of the type found in many gardens. Sunlight passes through the transparent glass without being absorbed and strikes the earth and other contents of the greenhouse, warming them up. The warmed objects then try to cool down by emitting heat radiation, but this has a much longer wavelength than that of light and cannot easily pass through the glass, which reflects most of the heat back into the greenhouse (cf. Figure 3.5[a]). This process continues until the glass has warmed up to the point where it radiates as much power outwards as that of the sunlight coming into it. The latter process is assisted by convection: air near the bottom of the greenhouse is heated, becoming less dense and rising to the top of the greenhouse, where it helps warm the glass as it cools and then falls back downwards.

Similar principles govern the greenhouse effect in the Earth's atmosphere (cf. Figure 3.5[b]). Sunlight passes through the atmosphere largely unhindered and warms the Earth's surface; the warmed surface radiates heat and some of this radiation is absorbed in the upper atmosphere and re-emitted, about half of the re-emitted energy returning to the Earth's surface. This is where quantum physics plays an important role. As I pointed out in Chapter 2 and have noted several times in this chapter, when electrons are confined within an atom or molecule, wave–particle duality ensures that the energy of the system must have one of a set of quantized values. Moreover, the excitation of such a system from its ground state can be caused by the absorption of a photon, but only if its energy matches the difference

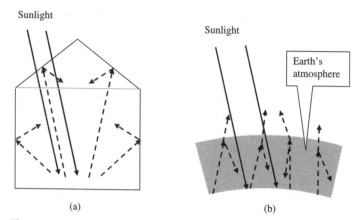

Sunlight

Sunlight

Earth's atmosphere

(a)   (b)

Figure 3.5 Sunlight (represented by solid lines) can pass through the glass of a greenhouse (a) and warm the contents, which emit heat radiation (broken lines). The wavelength of this radiation is much longer than that of the sunlight and cannot readily pass through the glass, so there is a net heating. A similar effect occurs on the Earth's surface (b). The Earth's atmosphere is transparent to sunlight, but contains carbon dioxide and other greenhouse gases that absorb heat radiation and re-radiate some of it back to the Earth's surface.

between the energies of the levels. Heat radiation has a wavelength of the order of 10^{-6} m and, as I explain in a little more detail below, the energy of such a photon is similar to the separation between the energy levels associated with vibration of the atoms within the molecule; such vibrations are not readily excited in molecules such as oxygen and nitrogen (the common constituents of air), but can be in others – in particular water and carbon dioxide. A photon that strikes one of these molecules can be absorbed, leaving the molecule in an excited state. It quickly returns to the ground state by emitting a photon, but this can be in any direction and it is just as likely to return towards the Earth as it is to be lost to outer space. These 'greenhouse gases' therefore play a similar role to the glass in the conventional greenhouse and this process leads to a heating of the Earth and its atmosphere until it is hot enough to re-radiate all the energy striking it. It is estimated that in the absence of carbon dioxide the temperature of the Earth's surface would be about twenty degrees Celsius less than it is today, while if the present amount of carbon dioxide in the atmosphere were to be doubled, the Earth's temperature would rise by between five and ten degrees Celsius, which would endanger the delicate balance on which life depends. As mentioned above, water is also an effective greenhouse gas, but the amount of water vapour in the atmosphere is determined by a balance between the evaporation of liquid water on the Earth, notably the surface of the oceans, and its re-condensation. This is controlled by the temperature of the Earth and its atmosphere and remains largely unchanged. However, if the Earth's temperature were to be raised substantially, the increase in atmospheric water vapour would be significant, which would lead to further global warming, then the production of more water vapour and so on. We would have runaway global warming of a type believed to have occurred on the planet Venus, where the surface temperature is now around 450°C.

However, in the short term at least, our concern is not with water vapour but with other gases, such as methane and especially carbon dioxide. As the amount of carbon dioxide in the atmosphere increases, the greenhouse effect leads to a corresponding rise in the Earth's temperature, an effect known as 'global warming'. Such an increase is today caused by human activity, particularly the burning of fossil fuels. The concentration of carbon dioxide in the atmosphere is estimated to have increased by about thirty per cent since industrial activity began in about 1700, and is currently increasing by about 0.5% per year, which, if it continued, would lead to a doubling in about 150 years and a consequent global warming of between five and ten degrees Celsius.

We can understand in a little more detail how quantum physics ensures that gases such as carbon dioxide act as greenhouse gases while the more common constituents of air – nitrogen and oxygen – do not. We saw in Chapter 2 that the energy required to excite an electron from the ground state of a typical atom corresponds to that of a photon associated with visible light. However, the energy of a photon associated with the heat radiated from the Earth's surface is about ten times less than this, so a different kind of process must be associated with the absorption of this low-energy radiation. The key point here is that in a molecule the atomic nuclei can be made to vibrate relative to each other. When we discussed the formation of molecules earlier in this chapter, we found that the distance between the nuclei corresponds to the point where the various contributions to the energy add up to the smallest possible total (Figure 3.1[b]). This means that if we were able to move the nuclei a little away from this equilibrium position, the energy would be raised, so that if we now released them the nuclei would move back towards the equilibrium point, converting the excess energy into kinetic energy associated with their motion. They would then overshoot the equilibrium point, slow down and return, and this vibrational motion would continue indefinitely unless the energy were lost in some way. In this sense a molecule behaves as if the nuclei were point masses connected by springs, undergoing oscillation as the springs stretch and contract. Figure 3.6 illustrates this for the case of the carbon dioxide molecule, which consists of a carbon atom bound to two oxygen atoms in a linear configuration. We discussed the quantum physics of an oscillator in Chapter 2, where we saw that the oscillator has a spectrum of energy levels, separated by Planck's constant times the classical oscillation frequency. We also saw in Chapter 2 that the energy of a quantum of radiation equals Planck's constant times the radiation frequency, so it follows that energy will be absorbed if this matches the oscillator frequency. Heat radiated from the Earth's surface has a range of frequencies, which encompass the vibration frequencies of the gases in the atmosphere, including those of greenhouse gases such as carbon dioxide.

The above applies just as much to nitrogen and oxygen as it does to carbon dioxide and water, so we still have to understand why heat radiation can induce vibration in the latter two gases but not the former two. To address this, we first recall the discussion of the wave function of the electrons in an atom in Chapter 2, where we noted that as long as the atom stayed in its ground state, we could think of the electronic charge as being diffused over the volume of the atom, with a concentration at any point that is proportional to the square of the wave function at that point. A similar situation applies to the ground state of a molecule; to a

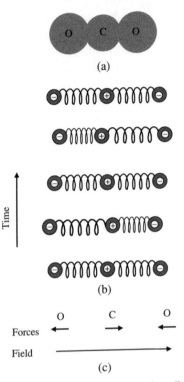

Figure 3.6 The electronic charge clouds in the carbon dioxide molecule are illustrated in (a). The atoms in a molecule can move as if they were connected by springs as illustrated in (b). In the carbon dioxide molecule, the central carbon carries a net positive electric charge and the two outer oxygen atoms are negatively charged. When an electric field is applied to the molecule, oppositely directed forces are applied to the oxygen and carbon atoms, which then respond as shown in (c), so exciting the vibration illustrated in (b).

first approximation the charge distribution has the form of over-lapping spherical clouds, as indicated in Figure 3.1(a) for the case of hydrogen. Because the two hydrogen atoms are identical, this molecule is symmetric and the two overlapping charge clouds are also identical. However, this is not true in the case of more complex molecules. Considering the lowest energy state of carbon dioxide in particular, it turns out that the total charge in the cloud surrounding the central carbon atom is a little less than six electronic charges and so does not fully balance the charge on the carbon nucleus, while the charge surrounding each oxygen is a little more than that corresponding to the total of eight associated with a free oxygen atom. The net effect of this is that, although the total electronic charge on the molecule balances the total nuclear charge, each oxygen atom carries a small net negative charge, and a balancing positive charge is associated with the carbon atom. We now consider what happens when the molecule is subjected to an electric field directed along its length. Returning to Figure 3.6(b), we see that this pulls the carbon in one direction and the oxygens in the other, so an electromagnetic wave that vibrates at the correct frequency can excite the molecule into a vibrational motion in which the carbon atom moves in the opposite direction to the two oxygen atoms. This allows the absorption of energy which is then re-emitted in a random direction, so leading to a greenhouse effect. A similar process occurs when an applied field is perpendicular to the line of the molecule: the carbon atom moves in one direction and the two oxygens in the other, an effect that in this case causes the molecule to bend. This also leads to a greenhouse effect for radiation of the appropriate frequency.

Why then does a similar effect not arise in a molecule like oxygen or nitrogen? The reason is that such a molecule contains two identical atoms, which must therefore either be neutral or carry the same net charge. In either case, they cannot be pushed in opposite directions by an electric field, so a vibration cannot be set up by an electromagnetic wave and such a gas cannot contribute to the greenhouse effect.

If quantum physics plays a role in creating the greenhouse effect and its associated problems, can it also help us avoid and resolve them? We already know that nuclear reactions are governed by the laws of quantum physics and produce no carbon dioxide or other greenhouse gases. Thus, the generation of nuclear energy (both fission and fusion) makes no contribution to the greenhouse effect. We have seen that nuclear energy may have problems of its own and it certainly has had a bad press since about 1980. However, some environmentalists have been revising their opinions in recent years. Other, 'green' forms of power include wind power, wave power and solar power. Although air and water are composed of atoms, which in turn depend on quantum physics for their existence and properties, the motion of wind and waves is governed by classical physics and is independent of the internal structure of the atoms, so, as we noted in Chapter 1, we do not classify this as a manifestation of quantum physics. Solar energy comes in two main forms. It can be used to heat domestic hot water systems (for example) and again there is nothing peculiarly quantum about this process, but it can also be used to produce electricity in 'photovoltaic cells', whose performance does depend on quantum effects. To understand this, we shall first have to find out how quantum physics can be applied to construct electronic devices. This is discussed in the next two chapters. We return to the operating principles of photovoltaic cells towards the end of Chapter 5.

summary

This chapter has discussed the role of quantum physics in the production of the energy we use to power our civilization. The main points are:

- When atoms join to form a molecule, the energy of the resulting ground state is less than that of the separated atoms and the surplus is given off as heat. This principle underlies all energy production by chemical combustion.
- When two nuclei fuse, energy is released in much the same way as in combustion except that the power produced is millions of times greater. Fusion is the process that fuels energy production by the sun and the stars and the explosion of a hydrogen bomb. Research into controlled fusion for energy production is ongoing.
- When a neutron is added to a heavy nucleus, such as U^{235}, it undergoes fission into smaller fragments. More neutrons are released in this process, which can lead to a chain reaction; this can be explosive, as in an atomic bomb, or can be controlled in a nuclear reactor to produce power.

- Quantum physics underlies the processes whereby global warming results from the greenhouse effect. Sunlight passes through the Earth's atmosphere and warms the Earth, but heat radiated from the Earth can be captured by carbon dioxide in the atmosphere, which re-radiates about half of it towards the Earth's surface. The amount of carbon dioxide in the atmosphere increases as a result of the burning of fossil fuels.

notes

1. The most promising process for controlled fusion is not the fusion of two protons to form deuterium, but the fusion of a deuterium nucleus with one of tritium (another isotope of hydrogen composed of one proton and two neutrons) to form a helium nucleus.
2. Joint European Torus: so called because of the doughnut shape of the machine.

metals and insulators

Those of us fortunate enough to experience it know how essential electricity is to modern living. It brings us the power we use to light our homes and our streets, to cook our food and to drive the computers that process our information. This chapter aims to explain how all this is a manifestation of the principles of quantum physics and, in particular, how quantum physics allows us to understand why the electrical properties of different solids can vary from metals that readily conduct electricity, to insulators that do not. Chapter 5 extends this discussion to 'semiconductors' – materials with the properties needed to allow us to construct the computer chips at the heart of our information technology.

First I must emphasize again that electricity is not a source of energy in itself, but rather a way of transmitting energy from one place to another. Electricity is generated in a power station which, in turn, gets its energy from some form of fuel (e.g. oil, gas or nuclear material) or perhaps from the wind, waves or sun. Quantum physics plays a part in some of these processes too, as we saw in Chapter 3.

Electricity comes to us in the form of an electric current that flows through a network of metal wires that stretch all the way from the power station, through the plug in the wall, to the computer that I am using to write this chapter (Figure 4.1[a]). A simple electric circuit is shown in Figure 4.1(b). This consists of a battery, which drives an electric current round a circuit that contains a resistor. We need to have some understanding of how this happens and what these terms mean. First, the battery: this

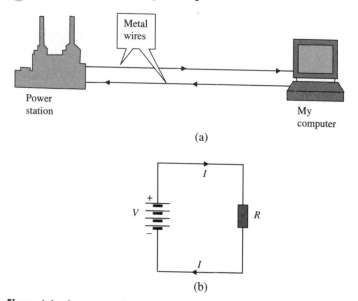

Figure 4.1 A power station driving an electric current through one wire to my computer and back along another is illustrated in (a). This requires that the wires are made of metal so that they can conduct electricity. A simple electric circuit is shown in (b). The voltage V created by a battery drives a current I round the circuit and through the resistor R. Because electrons are negatively charged they move in the opposite direction to that of the conventional current.

consists of a number of 'electrochemical cells' that use a chemical process to generate positive and negative electrical charges on opposite ends of each cell. These can then exert a force on any mobile charges connected to them and their potential for doing so is termed the 'voltage' produced by the battery. Next the connecting wires: these are made of metal and (as I shall discuss in some detail shortly) metals are materials that contain electrons which are able to move freely within the material. When a wire is connected to a battery as in Figure 4.1(b), electrons close to the negative terminal of the battery experience a repulsive force driving them through the wire; they travel around the circuit until they reach the positive terminal to which they are attracted; they then pass through the battery and emerge at the negative terminal where the process is repeated. As a result, a current flows around the circuit and we should note that, because the electrons carry a negative charge, the conventional direction of current flow is opposite to that of the electrons. The reason for this is simply that the concept of electric current was developed and the conventional meaning of positive and negative charge was established, before electrons were discovered. Figure 4.1(b) also shows the current passing through a resistor, which, as its name implies, is a device that 'resists' the flow of current; its ability to do so is determined by a property known as its 'resistance'. The voltage needed to drive a given current through a given resistor is proportional to the size of the current and to the resisitance; this relationship is known as 'Ohm's law', which we shall discuss in more detail towards the end of this chapter. Apart from superconductors (to be discussed in Chapter 5), all materials present some resistance to electrical current, though the resistance of a typical copper wire is very small. Resistors are often constructed from particular metal alloys, designed to present significant resistance to the flow of current; the current that does flow through them loses some of its energy, which is converted into heat. This is the process that underlies the operation of any electrical heater, such as may be found in a kettle or a washing machine or used to heat a room.

Some materials, such as glass, wood and most plastics, are 'insulators' whose resistance is so high that they essentially do not allow the passage of any electrical current. They play a vital role in the design of electrical circuits, because they can be used to separate current-carrying wires and so ensure that electrical currents flow where we want them. Readers are probably familiar with this in a domestic context, where the wires carrying electrical current around our houses are protected by plastic sheaths that prevent them connecting with each other or coming into contact with ourselves.

The difference between metals and insulators is one of the most dramatic of any known physical properties. A good metal is able to conduct electricity well over a trillion (10^{12}) times more efficiently than is a good insulator. Yet we know that all materials are composed of atoms that contain electrons and nuclei. How is it that their properties can be so different? Once again, we shall see that the answer lies in quantum physics: if the electrons were not quantum objects with wave properties, none of this would be possible.

In our discussion of atoms in Chapter 2, we found that the electrons occupied 'shells' of energy states, with often just one or two electrons in the outermost, highest-energy, shell. When such atoms form molecules these electrons are no longer bound to particular atoms, but can move freely between them; an example of this is the hydrogen molecule discussed in Chapter 3. A solid is in some ways like a giant molecule: the atoms are held quite close together and the outer electrons are no longer bound to particular atoms. From now on we shall distinguish between these 'free' electrons and the positively charged 'ions', by which we mean the nucleus of an atom along with its inner-shell electrons. As a first approximation, we shall assume that the behaviour of the free electrons is unaffected by the ions. Later we shall see how this picture has to be modified by the presence of the ions and we shall find that if the ions form a regular array, as they do in a crystal, the wave properties of the electrons mean that their motion is largely unhindered in metals, but totally obstructed in insulators.

We saw above that electric currents move in circuits that generally include a source of energy (the battery) driving the current and a load (the resistor). We shall simplify this and consider just a loop of wire as in Figure 4.2 and, as we are ignoring the ions, we can assume that the potential energy the electron experiences is the same at all points on the loop. Electrons are able to flow round this loop in both directions, so if a current is flowing, more electrons must be moving one way than the other, while if no current flows, the same number of electrons are moving clockwise and anticlockwise. Consider a particular electron that moves in one direction at some speed. From what we have learned in earlier chapters, the wave function associated with the electron has the form of a travelling wave, whose wavelength is determined by the electron's speed. However, because the metal is in the form of a closed loop, such a wave can exist only if it joins up with itself after going right round the loop, which means that the total distance round the loop must equal a whole number of wavelengths, as shown in Figure 4.2 and explained in more detail in Mathematical Box 4.1. We note that as the electron can move in either

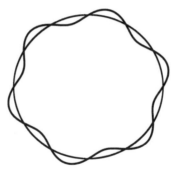

Figure 4.2 A loop of wire forming an electric circuit. A whole number of wavelengths of the electron waves must match the distance round the circle.

direction, two travelling waves correspond to each allowed value of the wavelength and that, because of spin and the exclusion principle discussed towards the end of Chapter 2, each of these waves can accommodate two electrons; each energy state can therefore contain up to four electrons. To simplify our discussion, we shall assume that the wire is very thin so that the electron motion is confined to that around the loop and we neglect motion across the wire. Most of the physical properties of this one-dimensional model turn out to be very similar to those of real three-dimensional solids. There are some exceptions to this rule, however, and we shall return to these later in this chapter.

We saw in Chapter 2 that the shorter the wavelength, the larger is the electron energy. In the absence of excitation, the electrons will fill the available states, starting with those of lowest energy, with four electrons associated with each allowed value of the wavelength. A classical analogy might be placing a number of balls into a bucket: the first few we put in will occupy the positions of lowest potential energy near the bottom of the bucket, but later on this space will be occupied and further balls will have to go into states of higher energy. In the quantum case, the energy of the filled state of maximum energy is known as the 'Fermi energy', after the Italian physicist Enrico Fermi. We should note that the total number of electrons is very large: over ten billion for a chain

MATHEMATICAL BOX 4.1

Referring to Figure 4.2, we assume that the total length of the wire is L and that the electron wavelength is l. It follows that the electron wave will join up with itself after going right round the loop if and only if

$$L = nl$$

where n is a whole number. Following exactly the same argument as in the case of the particle in a box (see Mathematical Box 2.5), we see that this implies that the electron energy equals E_n, where

$$E_n = (h^2/2mL^2)n^2$$

There are two travelling waves (one clockwise and one anticlockwise) corresponding to each value of n. Each of these waves can have up to two electrons (with opposite spin) associated with it. Thus the first four electrons will occupy the energy level with $l = L$ and energy $h^2/2mL^2$, the next four that with $l = L/2$ and energy $4h^2/8mL^2$, the next four that with $l = L/3$ and energy $9h^2/8mL^2$ and so on. Hence, if the metal contains N electrons, electrons will occupy the $N/4$ levels of longest wavelength so that the wavelength of the highest-energy filled level is $4L/N$, implying that its energy equals $(h^2/8mL^2)N^2$. (Note that this explicitly assumes that N is divisible by 4; if this is not the case, the highest level will contain fewer than four electrons.)

of atoms ten centimetres long. However, as we shall see, the states that are relevant to electrical conduction are those at or close to the Fermi energy; these are illustrated in Figure 4.3(a).

Now consider what will happen if we try to pass a current through the metal by applying a voltage that exerts a force on the electrons in (say) the clockwise direction (assumed to be from left to right in Figure 4.3), which is essentially what happens if we include a battery or other power source in our circuit. This force tends to increase the speed of an electron that is already moving from left to right, and decrease that of one moving in the opposite direction. However, a change in the electron speed implies a change in its momentum, and hence its wavelength and therefore

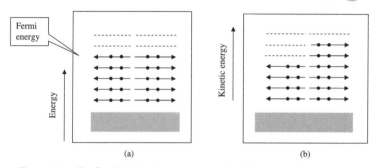

Figure 4.3 The five highest electron energy levels in a metal. An arrow with two dots pointing from left to right represents a wave that travels clockwise around a circuit like that shown in Figure 4.2, and contains two electrons with opposite spin. The broken lines represent unoccupied energy levels and the grey areas indicate a large number of energy levels that are not explicitly shown. The filled dots indicate electrons and their positions on the lines have no significance. In (a) the electrons fill clockwise and anticlockwise states equally and no net current flows. In (b) an external force has increased the speed and hence the kinetic energy of the electrons moving from left to right (clockwise) and decreased that of the electrons moving from right to left (anticlockwise). As a result, there are now more filled clockwise than anticlockwise waves and a net anticlockwise current.

its quantum state, which is possible only if allowed by the exclusion principle. The net result is that the balance between clockwise and anticlockwise electrons is unaffected except for those electrons whose kinetic energies are at, or close to, the Fermi energy. In the case of these, the effect of the voltage is to transfer some anticlockwise electrons with energies just below the Fermi energy to previously empty clockwise states just above it, leading to a net clockwise flow of electrons and hence an electric current as illustrated in Figure 4.3(b). As discussed earlier, the size of this current is controlled by the voltage and the resistance of the material. We shall return to the reasons for this resistance later in this chapter.

Although I have simplified things somewhat, the above description is pretty much what happens when an electric current flows round a metal circuit. However, we have still to understand why some materials behave like this whereas others act as insulators, preventing all current flow. To address this point, we shall have to consider the role of the ions.

what about the ions?

So far we have assumed that the electrons are free to move anywhere in the metal unhindered, but we know that all solids are composed of atoms. We can reasonably expect that one or more electrons in the highest energy shell will be able to move easily from one atom to the other, and as a first approximation we have ignored any interaction between them and the ions. However, it is hard to see how we can justify ignoring the ions completely, since the latter carry a net positive charge, which should interact strongly with the negatively charged electrons. We might therefore expect that an electron attempting to move through the metal would undergo a series of collisions with the ions, which would greatly impede its motion, preventing significant current flow. As an analogy, consider trying to walk straight through a dense forest: bumping into trees would continually hamper your progress, slowing you down or possibly completely preventing your progress. Why does something similar not happen to the electrons? There are two reasons for this: the first is down to quantum physics and the fact that the electrons have wave properties; the second is that it is a key feature of solids that their ions are arranged in a regular, periodic pattern, because they are made up of crystals. We shall now see how these features combine to determine the electrical properties of solids.

We are all familiar with crystals, though we may think of them as rather exotic objects, such as expensive gemstones (Figure 4.4[a]) or crystals carefully grown in a school science lesson. It may come as a surprise to learn that many solids, including metals, are crystalline. We shall return to this point shortly, but first we'll look at some of the main properties of crystals and how these are reflected in their atomic structure. Crystals are noted for their flat faces, sharp edges and regular shapes. Moreover, if a crystal is cut into one or more bits, these properties – in particular the regular shape – are preserved. When the atomic composition

(a)

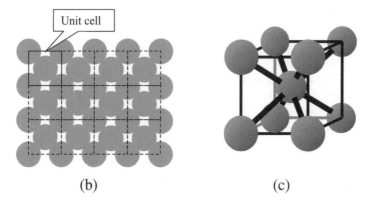

(b)   (c)

Figure 4.4 Crystals and their atomic structures. A single crystal of diamond is shown in (a). In (b) we see how disks can be arranged to form a square pattern that repeats itself across the plane of the page. Similarly, in three dimensions spheres can pack together to form a cubic unit cell and this is illustrated in (c) (where, for clarity, touching spheres are replaced by smaller spheres connected by rods). An example of this structure is that of a crystal of copper.

of matter was revealed in the nineteenth century, this led to the discovery that crystal shapes are a consequence of regularities in their atomic structure. In other words, crystals are composed of a huge number of identical building blocks of atomic dimensions. A simple and common example of such a building block, or 'unit cell' as it is now known, is a cube. As shown in Figure 4.4, atoms can be arranged to form cubic unit cells, and a given crystal is built up from a continued repetition of identical cells. Figure

4.4(b) illustrates this for circles packed together in a plane and Figure 4.4(c) shows how the unit cell of copper is built up from copper atoms. Confirmation of atomic structure was obtained in the twentieth century when crystals were probed using the recently discovered X-rays. The patterns formed by the scattered X-rays were just those expected from a periodic array of objects with atomic dimensions. These experiments also revealed how ubiquitous crystalline properties are. Although many materials have a crystalline structure, this is not always immediately obvious because often a sample does not consist of single crystal but is composed of a large number of randomly oriented crystalline 'grains'. These grains, with a typical size of one micrometre (10^{-6} m) are small on an everyday scale, but about a thousand times larger than a typical atom. We will assume that a solid consists of a single crystal and that if we can understand how a metal crystal is able to conduct electricity, it is reasonable to assume that current can flow from one grain to another in a typical wire; this assumption is well confirmed experimentally.

The property of a crystal that provides an explanation for the electrical conduction of metals and the insulating properties of many other materials is its regular repeating structure. We know that a wave is also something that repeats in space and it turns out that the interaction between a wave and a crystal is very weak unless there is a match between the crystal spacing and the wavelength of the wave, in which case it is very strong. We shall now try to understand this in a little more detail and see how it produces the effects we have discussed. So far when discussing the motion of electrons round our loop, we have assumed that the potential energy inside the loop of wire is everywhere the same. We will now consider what happens if the loop contains ions arranged in a periodic array, so forming a chain of regularly spaced ions, which we can think of as a one-dimensional crystal. We know that the electrons are negatively charged, while the ions carry a positive charge, so the electrostatic potential is smaller (i.e. more attractive) in the vicinity of the ions than it is, say, halfway between them and we might expect this to have a significant effect on the values of the allowed energy levels and the form of the associated wave functions. Given that a fundamental property of the wave function is that its square equals the probability of

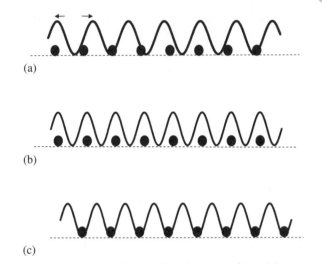

(a)

(b)

(c)

Figure 4.5 The interaction between the electrons and ions in a one-dimensional crystal. According to quantum physics, the electrons are most likely to be found where the squared wave function (represented by a curve) is largest. The black circles mark the positions of the ions. In (a) there is no match between the wavelength of the wave and the separation between the ions, and the overall interaction energy does not depend on the position of the wave relative to the ions. In (b) the ions are in regions where the electrons are most likely to be found, so the interaction energy is large and negative. In (c) the ions are in regions unlikely to be occupied by electrons, so the interaction energy is small.

finding a particle at any point, Figure 4.5 shows the squared wave function in the case of an electron in a wire, along with the ions, and for several values of the wavelength. In most cases, such as that shown in Figure 4.5(a), some ions are at points where the probability of finding an electron is high, some at points where it is low and some in between. The average value of the potential energy of interaction is then just equal to the average of the potential at all points on the loop. We therefore conclude that the energies of all such states are reduced by the same amount, which has no effect on their relative energies, so the earlier arguments based on completely ignoring the ions (cf. Figure 4.3[a]) are largely unaffected.

However, when the wave and the ions are in step so that the electron wavelength equals twice the repeat distance separating the ions, as in Figure 4.5(b) and (c), the interaction between the wave and the crystal of ions is quite different. At one extreme (illustrated in Figure 4.5[b]) the wave function has the form of a standing wave with its maximum intensity at the ionic positions. These are the points where the electron is most likely to be found and, when it is in this vicinity the energy is lowered because of the attraction between the negatively charged electron and the positively charged ion. At the other extreme (Figure 4.5[c]) the standing wave is placed so that the electrons are most probably found halfway between the ions, so that the energy is raised above the average instead of being lowered. Thus the wave functions representing these particular states have the form of standing waves with energies either significantly greater or smaller than the average. A full solution to the Schrödinger equation confirms this picture and shows that standing waves of this wavelength are always locked into the ionic positions in one or other of the configurations just described. Moreover, it turns out that waves with wavelengths close to but greater than twice the repeat distance also have their potential energies reduced somewhat, while those with slightly shorter wavelengths have their potential energies increased. As a result, a gap appears in the spectrum of energy levels, as illustrated in Figure 4.6. Note that the energies of the states well below the gap are essentially unaffected by the presence of the ions, for the reasons discussed earlier.

This energy gap plays a key role in determining the contrasting properties of metals and insulators. We saw earlier that, in general, only those electrons that occupy states with energies close to the Fermi energy contribute to current flow. If this corresponds to a state well below the energy gap, current will flow in the same way as discussed earlier when we ignored the ions. However, if the maximum occupied energy level is that just below the energy gap, the imbalance illustrated in Figure 4.3(b) cannot develop because the required empty states are not accessible; this is demonstrated in Figure 4.6. These points are developed further in Mathematical Box 4.2, where we derive the simple result that solids with an odd number of electrons per atom are predicted to be metals, whereas those possessing an even number per atom should be insulators.

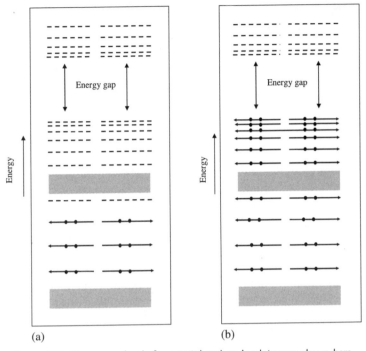

Figure 4.6 The energy levels for a metal and an insulator are shown here. As in Figure 4.3 the arrows represent clockwise and anticlockwise waves that are occupied by electrons (small filled circles) and the broken lines correspond to empty states. The positions of the symbols on the lines have no significance. The grey areas indicate bands of energy levels that are not shown individually. The interaction between the electron waves and the crystal of ions leads to a gap in the band of energy levels as shown. In a metal (a), there are only enough electrons to half-fill the band, and there are no occupied states in the vicinity of the gap; current can therefore flow in the same way as is illustrated in Figure 4.3(b). In contrast, an insulator (b) contains exactly the right number of electrons required to fill the band; so there are no accessible empty states and current flow is impossible.

How does this simple rule predicting whether a substance is a metal or an insulator hold up in practice? If we consult the periodic table of the elements, we find that the best conductors do indeed have an odd number of electrons: for example the alkali metals, lithium (three electrons), sodium (eleven), potassium

MATHEMATICAL BOX 4.2

Let us consider again the electron waves in a one-dimensional metal loop of length L, as illustrated in Figure 4.2. We saw in Mathematical Box 4.1 that the allowed values of the electron wavelengths are

$$l_n = L/n$$

where L is the length of the piece of metal we are considering and n is a whole number. Two possible waves (one clockwise and one anticlockwise) correspond to each of these wavelengths and up to two electrons (with opposite spin) can be associated with each travelling wave, so that each value of n corresponds to four electron states. Now let us suppose that the metal is a one-dimensional crystal containing N atoms, the separation between neighbours being a, so that $L = Na$. It follows from the discussion in the text that the energy gap corresponds to waves with wavelength $2a$. If we now assume that each atom possesses a single electron that is loosely bound to it and can be expected to become free in the metal, these N electrons will fill the $N/4$ states of lowest energy, which must therefore all have wavelength longer than $l_{N/4}$, where

$$l_{N/4} = 4L/N = 4Na/N = 4a$$

This means that we have only half the electrons needed to fill the band of energy states below the gap and none of the states near the gap contains electrons. We can conclude that the electrons in a one-dimensional solid made up from atoms with one electron in their outer shells will be unaffected by the ions and the substance will be a metal.

In contrast, consider what happens if the atom has two electrons in its outer shell. All states with wavelengths longer than $l_{N/2}$, where

$$l_{N/2} = 2L/N = 2Na/N = 2a$$

will now be full, which means that there are now just enough electrons to fill all the states in the band right up to the gap. We saw earlier that to get a net current, there must be an imbalance in the number of electrons associated with waves travelling in opposite directions and that in metals this is achieved by exciting some electrons into previously empty states of energy a little greater than the Fermi energy in the absence of a current. However, this is now impossible, since all such states are separated from the occupied states by the energy gap, which is generally too large to be overcome by the energy supplied from the voltage source that is attempting to drive a current. We conclude that a solid containing two free electrons per atom is incapable of supporting a current and is therefore an insulator.[1] If we now consider the case where there are three electrons per atom, the electrons will occupy states up to an energy corresponding to a wavelength of $2a/3$, which is halfway between the top of the first energy gap and the bottom of the second. We therefore expect this material also to be a metal. However, for four electrons per atom the minimum wavelength is a, which is halfway between the top of the lowest and the foot of the second-lowest energy gap, so the material is an insulator. Extending this argument, we conclude that atoms with an odd number of electrons should be metals whereas those with an even number should be insulators.

(nineteen); and the 'noble metals', copper (twenty-nine), silver (forty-seven) and gold (seventy-nine). Some quite common metals obey this rule, though this is not obvious at first sight: for example, iron has twenty-six electrons in total, but it is known that only one of these is in an outer shell, where it is relatively easily freed from the atom; the rest are tightly bound in inner low-energy shells. Examples of non-metals with an even number of electrons are carbon (six), silicon (fourteen) and sulphur (sixteen). High-quality insulators are nearly always built up from molecules rather than single atoms: for example, paraffin wax consists mainly of 'straight-chain hydrocarbons' such as $H_3C(CH_2)_nCH_3$ which contain a total of $8n + 18$ electrons (where, as always, n is an integer) – which is always an even number.

There are, however, notable exceptions to our general rule. For example, calcium (twenty) and strontium (thirty-eight) are known to be metallic, despite having two electrons in their outer shells. It turns out that this is due to the fact that the real world has three dimensions, whereas we have based our argument on a simple one-dimensional model. When quantum theory is applied to realistic three-dimensional solids, it agrees with the one-dimensional rule that solids whose basic building blocks (be they atoms or molecules) contain an odd number of weakly bound electrons should always be metals, but the situation is less clear cut if the number of weakly bound electrons is even. We then have to consider waves moving in different directions – at an angle to the current flow rather than directly along it. It can be shown that in some circumstances this creates states in the upper band that have lower energy than some other states in the lower band. When electrons are transferred from the lower to the upper band to reduce the overall energy, both bands end up only partly filled and therefore able to support current flow.

These complications only reinforce the fact that metals' ability to conduct electricity depends critically on the ability of the electrons to pass largely undisturbed through the crystal formed by the ions. They can do this only because they have wave properties, which, in turn, depend on quantum physics. The dramatic difference between metals and good insulators occurs because of the full bands and large energy gaps in the latter materials and the only partly filled bands in the former.

a bit more about metals

The fact that the energy of an electron in a half-filled band is independent of the wavelength of the wave means that the wave function of an electron in a metal can have the form of a travelling wave that can pass through a crystal lattice without interacting with it. If this were the whole story, metals would be perfect conductors of electricity. However, in practice, although metals conduct electricity very well, they still present some resistance to current flow. The reason for this is that a crystal's periodic atomic structure is never completely perfect and the imperfections obstruct the current flow. Two main types of imperfection are commonly encountered. The first we shall discuss are impurities –

i.e. atoms of a different type from the main constituent of the material. Typically, these will be distributed more or less randomly through the crystal, upsetting its periodicity at these points. The second imperfection arises because the ions are constantly moving due to the effects of temperature: at any moment some of them will be a significant distance away from their standard positions so that again the crystal's periodicity is disturbed. The net result of all this is that, although electrons in metals do pass quite freely through the crystal for the reasons discussed above, they are scattered from time to time by impurities and by thermal defects. Typically an electron travels a distance of a few hundred ion spacings before meeting an impurity or thermally displaced ion, but when it does interact with such a defect, it loses its forward momentum and moves away in a random direction. At the same time, the electric force acts on the electron to push it forward again in the direction of the current flow. There is therefore a competition between the electric force pushing the electrons forward and an effective force associated with the defect scattering which tries to resist this. As a result, the size of the current flowing through a particular sample increases in proportion to the electric field and hence the voltage applied. This result is a familiar property of electrical conduction which is known as 'Ohm's law' – see Mathematical Box 4.3. Moreover, the number and size of the defects that provide the scattering determine the magnitude of the resistance to the current. For reasonably pure samples at room temperature those associated with thermal motion are usually the most important and their effect increases as the temperature rises, which leads to the well-known result that electrical resistance is proportional to absolute temperature. For some applications, such as electric heaters, we want to engineer materials that resist current flow quite strongly. This can be done by alloying two metals, which can be thought of as introducing a large proportion of impurity atoms of one type into a periodic crystal lattice of another. This high density of impurities then scatters the electrons quite strongly and produces a resistance that is largely independent of temperature.

In this chapter we have seen how quantum physics is essential to understanding the properties of metals and insulators and why

MATHEMATICAL BOX 4.3

When we apply an electric voltage of magnitude V across a piece of metal of length L, we create an electric field inside it and the size of this field is F, where

$$F = V/L$$

The force on an electron owing to this field is $-eF$, which accelerates the electron (mass m and charge e) so that in time t its velocity in the direction of the field increases by

$$v = -eFt/m = -eVt/mL$$

We suppose that, on average, an electron collides with a defect and rebounds in a random direction after a time t_0. Many electrons in the metal will follow this process, so it is reasonable to assume that the average speed after such a collision is zero. The field will then accelerate the electrons again and their velocities will increase back to the above value before being scattered again. If we average over a large number of such processes, we find that the average speed of an electron in the current direction is

$$v_{av} = -eVt_0/2m$$

An electron moving with velocity v contributes an amount $-ev$ to the electric current, so if there are n electrons in the metal the total current will be

$$I = -nev_{av} = (ne^2t_0/2mL)V$$

Thus the electrical resistance, R, is

$$R = V/I = 2mL/ne^2t_0$$

This is consistent with Ohm's law, which states that the resistance, R, defined as V/I, is constant. We see that the size of R is related to the time between collisions, t. This becomes smaller (making R larger) if the thermal motion increases because of a rise in temperature or if we increase the level of impurities in the metal.

they differ so markedly. In the next chapter, we shall discuss semi-conductors and see how quantum physics plays a role in determining their properties, which are essential to the information technology that plays such a large part in modern life.

summary

This chapter has explained how the principles of quantum physics underlie the electrical properties of solids – in particular why some solids are metals that conduct electricity, whereas others are insulators that obstruct the flow of electric current. The main points made are:

- In a solid, the electron waves span the whole sample to form a large number of very closely spaced energy levels, each of which can contain up to four electrons.
- In the absence of a net electric current, equal numbers of electrons move in opposite directions, but this balance is upset when an electric field is applied, provided there are accessible empty states available to the electrons near the Fermi energy.
- Many solids, particularly metals, are formed from crystals where the atoms form a regular, repeating pattern.
- The electrons are significantly affected by the array of ions only if their repeat distance matches the wavelength of the electron wave, in which case a gap appears in the spectrum of energy levels.
- In metals, there are only enough electrons to half-fill the band of states below the gap, which means that there are accessible empty states and hence current can flow.
- In insulators, the band is filled right up to the gap, so there are no accessible empty states and current flow is impossible.
- Our one-dimensional model predicts that solids with an odd or even number of electrons per atom will be metals and insulators respectively. There are exceptions to this rule in the real three-dimensional world.
- The flow of electrons through a metal is obstructed by collisions with thermal excitations and impurities, leading to electrical resistance and Ohm's law.

note

1.   If a large enough voltage is applied to an insulator, electrons
      can be forced into the upper band, causing a current to flow.
      This process is known as 'dielectric breakdown'.

semiconductors and computer chips

In the last chapter, we saw how the dramatic difference between metals and insulators was a consequence of the interaction between the waves associated with the electrons and the periodic array of atoms in the crystal. As a result, the allowed energies of the electrons form a set of bands, separated by gaps. If the solid contains enough electrons to just fill one or more bands, they then cannot respond to an electric field, and the material is normally an insulator. In contrast, the highest occupied energy band in a typical metal is only half full and the electrons readily respond to an applied field, producing a current flow.

This chapter describes the properties of a class of materials that lie between metals and insulators and are known as 'semiconductors'. Like insulators, semiconductors have an even number of electrons per atom, and therefore just enough to completely fill a number of bands. The difference is that in semiconductors the size of the gap between the top of the highest full band and the foot of the next empty band is fairly small – which means not very much larger than the energy associated with the thermal motion of an electron at room temperature. There is then a significant probability of some electrons being thermally excited from the full band into the empty band (see Figure 5.1). This has two consequences for electrical conduction. First, the electrons excited into the upper band (known as the 'conduction band') can move freely through the metal, carrying electric current, since there are plenty of available empty states to move into. Secondly, the empty states left behind in the lower band (known as the 'valence band') are

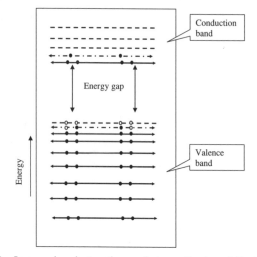

Figure 5.1 In a semiconductor, the gap between the top of the highest full band and the foot of the lowest empty band is small enough for some electrons to be thermally excited across the gap. This is illustrated in the above diagram, where continuous lines represent completely filled states (i.e. each contains four electrons indicated by filled circles), broken lines indicate empty states and dot-dash lines indicate partially filled states. Current can be carried both by the excited electrons and by the empty states or 'holes' (open circles) left in the valence band. The positions of the symbols on the lines have no significance.

available to the electrons in this band, so these are also able to move freely and carry current. Thus, both bands contribute to current flow and the material is no longer a perfect insulator.

We now consider the behaviour of the nearly full lower band in a little more detail. We will find that its properties are just the same as those of a nearly empty band, containing *positively* charged particles rather than negative electrons. To see how this works, we first recall that in a full band an equal number of electrons move in opposite directions. Referring to Figure 5.2(a), we see that if one of these electrons is removed, an imbalance results and the net result is a current equal but opposite to that associated with the missing electron; however, this is just the current that would result from a single positive charge moving with the same velocity as that of the missing electron. In Figure 5.2(b) we

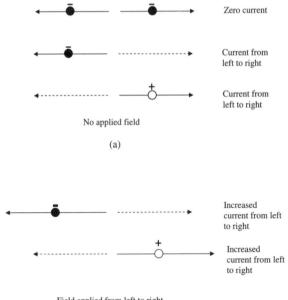

Figure 5.2 The effect of removing one of a pair of electrons moving with opposite speed is to create a state with the same properties as a positive charge moving in the same direction as the missing negative charge. In (a) we see that the electric current created by removing one electron from an oppositely moving pair creates a current flow that is equivalent to that of a positive charge with the same velocity as the missing electron. In (b) we see that applying a field from left to right increases the speed of the negative electron and hence the current; applying the same field to a positive charge would produce the same charge in current.

illustrate the effect of applying an electric field to the system: this exerts the same force on all the electrons, causing their velocities to change by the same amount, so the net change is again equal and opposite to what would have been experienced by the missing electron. Thus, the behaviour of a set of electrons with one removed is the same as that expected from a particle that possesses all the properties of a single electron, except that its electrical charge is positive.

Deeper study confirms that all the relevant properties of a nearly full band are identical to those of a nearly empty band containing a number of positively charged particles equal to the number of the missing electrons. As it is much easier to envisage the behaviour of a small number of positively charged particles than it is to consider the properties of a huge number of electrons, we will follow usual practice and employ this model from now on. These fictitious positive particles are conventionally known as 'holes', for reasons that should be pretty obvious. However, we should not forget that this is a convenient model and does not imply the presence of any real positive carriers of charge in the metal; we should also note that such 'holes' have nothing whatsoever to do with 'black holes'!

Returning to the particular example of a semiconductor (Figure 5.1), we can describe it as containing two bands that possess charged particles capable of transporting electric current. One of these is the upper band containing a number of free electrons and the other is the lower band that contains the same number of holes. The actual number of free electrons/holes depends on the temperature. In the case of pure silicon at room temperature, this number is equivalent to only about one in a trillion (10^{12}) of the total number of electrons in the lower band. Thus the total number of electrons and holes is very much smaller than the total number of electrons capable of carrying current in a typical metal. A semiconductor is therefore a much less efficient current carrier than a metal, though much more efficient than a typical insulator – hence its name.

We saw above that free electrons and holes can be created in a semiconductor because the energy gap is small enough to allow thermal excitation across it. As each of the electrons that have been promoted leaves a hole behind in the lower band, it is clear that this process generates equal numbers of electrons and holes. Researchers soon realized that materials where the free particles were either predominantly electrons or predominantly holes could have great advantages, and that this could be achieved by replacing some of the atoms of the material by other types of atom that contribute either *more* electrons to the solid, which would become the dominant charge carriers, or *fewer*, which would result in an excess of holes. In the case of silicon, a suitable element of the first type is phosphorus, since each phosphorus atom carries five electrons in its outer shell, compared with the four associated with each atom of silicon. When phosphorus impurities are introduced into a crystal of silicon, the additional electrons occupy energy states that are weakly bound to the phosphorus atoms and have energies within the silicon energy gap, a little below the bottom of the upper energy band. Because the energy difference between this level and the foot of the upper band is smaller than the band gap, a significant fraction of these electrons are excited into the upper band at room temperature, as illustrated in Figure 5.3(a). Boron, on the other hand, has only three electrons in its outer shell. When this type of atom is added to silicon, empty states are

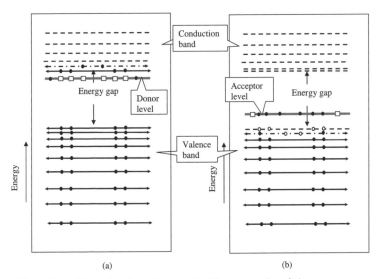

Figure 5.3 When phosphorus is added to silicon, a surplus of electrons are produced, which occupy a 'donor level' near the top of the energy gap. Electrons are readily excited thermally from this donor level to the conduction band and an n-type semiconductor is formed (a). Adding boron to silicon leads to a deficiency of electrons and the formation of an 'acceptor level'. Electrons are readily excited thermally from the valence band to the acceptor levels, leaving holes in the valence band and so creating a p-type semiconductor (b). Electrons are represented by filled circles, holes by open circles and vacant states in the donor and acceptor levels by open squares. The positions of the symbols on the lines have no significance.

created at an energy just above that of the top of the lower band: electrons are then thermally excited into these states from the lower band, creating a surplus of holes (Figure 5.3[b]). Semiconductors in which the charge carriers are predominantly (negative) electrons are known as 'n-type' semiconductors, whereas those whose properties are primarily determined by positive holes are known as 'p-type'.

In a typical device, the fraction of impurity atoms is about ten per million. About ten in every million impurity atoms has lost (or gained) an electron. Thus there is around one free electron for every ten billion atoms, which is about one hundred times the concentration of free electrons or holes in pure silicon at room temperature (see above).

An important step in the development of semiconductor technology was the discovery of practical methods to control the levels of impurity in these substances to a high level of precision and to allow materials to be made with a known number of positive or negative charge carriers. A typical impurity concentration of phosphorus or boron in silicon is about ten atoms per million, but in order to construct useful practical devices, we have to know and be able to control the impurity levels to an accuracy of one part in a million or better. This was one of the technological barriers that prevented the practical application of semiconductors for about twenty years in the middle of the twentieth century. However, these barriers were eventually overcome and today the semiconductor industry can produce semiconductors with purities controlled to one part per billion or better.

Before we can discuss some of the important applications of semiconductors, we will need to know what happens if more electrons or holes are introduced into a semiconductor than is expected at that temperature. As an example, consider the case of 'injecting' holes into an n-type semiconductor. The result of this is to create extra vacancies in the lower band; electrons then drop into these from the upper band until a new thermal equilibrium is reached. This process is described as the 'annihilation' of holes by electrons (or vice versa). As we shall see, it is important for the successful operation of transistors that this process is not instantaneous and that the holes can survive annihilation for a short time.

the p–n junction

One of the simplest devices to exploit the above properties is made by joining a piece of p-type semiconductor to one of n-type to form a 'p–n junction'. It is found that such a junction acts as a current rectifier, which means it is a good electrical conductor if current flows in one direction (from the n to the p), but presents a large resistance to current flow in the opposite direction.

To understand how this comes about, we first consider the junction region where the n-type and p-type meet. An electron entering this region may make a transition into a vacant level in the lower band, which thereby removes the electron along with a hole; we say that the electron and hole have been annihilated, as discussed above. As a result, there is a deficiency of both electrons and holes in the junction region and the charges on the ions are not fully cancelled out, leaving narrow bands of positive and negative charge on the n-type and p-type sides of the junction, respectively; these charges are known as 'space charges'. Referring to Figure 5.4(a), consider applying a voltage that would drive a current from p to n: in the p-type material the holes move in the direction of this current, while the electrons in the n-type move in the opposite direction. This leads to an increase in the number of both charge carriers in the junction region and a consequent reduction in the size of the space charges. When the electrons cross the centre of the junction on to the p-type material, they annihilate some of the holes, and electrons are similarly annihilated when holes pass into the n-type material. As the electrons and holes disappear from the n-type and p-type material, respectively, they are replaced by others drawn in from the external circuit, and a current flows. When a voltage is applied so as to drive a current in this direction, the junction is described as being 'forward biased'. In contrast, if a voltage is applied in a direction tending to drive a current from n to p (Figure 5.4(b)), the electrons and holes are drawn away from the central region; as a result, the space charges are increased and it becomes more difficult for charge carriers to cross the junction. Current cannot then continue to flow and in this configuration the junction is said to be 'reverse biased'. Thus, a current flows when a voltage is applied

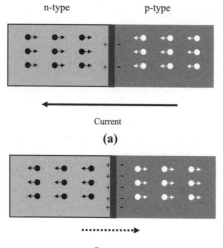

Figure 5.4 When current flows through a p–n junction from p to n, as in (a), electrons (filled circles with arrows denoting their direction of motion) and holes (open circles) are driven towards the centre of the junction, where they partly are annihilated, enabling current to flow through it. When we attempt to drive a current from n to p, as in (b), electrons and holes are pulled out of the junction region, so that the space charge in this area is built up, which prevents the flow of current.

in one direction but not when it is applied in the other, which means that a p–n junction has just the rectifying properties mentioned above.

Rectifiers such as those made from p–n junctions have many applications in the domestic and industrial applications of electricity. Electricity is produced in a power station by a generator driven by a rotating motor. One consequence of this is that the electricity produced is 'AC', which means that the voltage produced alternates from positive to negative and back every time the motor rotates – typically fifty times per second. Such alternating currents are perfectly satisfactory for many applications, such as room heaters and the electric motors used in washing machines

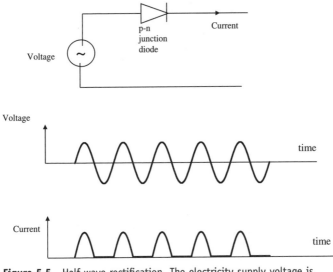

Figure 5.5 Half-wave rectification. The electricity supply voltage is normally AC, meaning that it oscillates in time, being positive for one half of the cycle and negative for the other. If this is applied to a p–n junction diode, current passes only during the half cycle when the voltage is positive.

etc. However, some applications require a power source where the current always flows in the same direction; this is essential, for example when we charge a battery in a car or use a charger supplied with a mobile phone. Rectifiers based on p–n junctions can be used to convert AC to DC (direct current) and this is illustrated in Figure 5.5, where we see that when an alternating voltage is applied to a p–n junction, current flows during only half of the cycle, so the resulting output will have only one sign, although it is zero for half the AC cycle. If we use four rectifiers connected as in Figure 5.6, the output voltage again has the same sign at all times, but is now on throughout the cycle. This is known as 'full-wave rectification' in contrast to the 'half-wave rectification' achieved by a single rectifier. Rectifiers based on p–n junction devices are part of all battery chargers. Sometimes a steady DC source is required: this may be provided by a charged battery or can be produced by

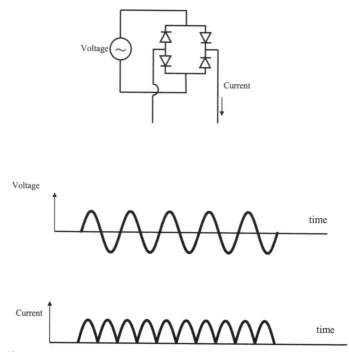

Figure 5.6   Full-wave rectification. When an AC voltage is applied to a set of four diodes arranged as shown, the output current will pass down the right-hand path when the voltage is positive and up the left-hand path when it is negative. As a result, the current has the same sign at all stages in the AC cycle.

smoothing the output from a rectifier using an electrical component known as a 'capacitor'.

the transistor

Essentially all modern information and communication technology is based on the semiconducting properties of silicon. Especially important is its use as a 'transistor'. A transistor is a device that can be used to turn a small signal, such as that detected by a radio receiver, into one of a similar form but powerful enough to

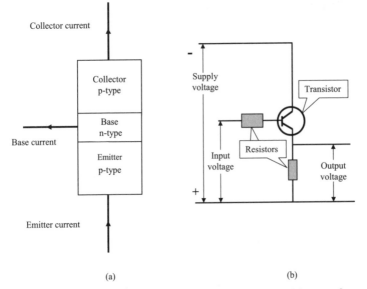

(a)   (b)

Figure 5.7 The operation of a p–n–p transistor is shown in (a). Part of the base current flows through the emitter and part is annihilated by holes passing through the base. As a result, a small change in the base current produces a large change in the collector current. This current gain is discussed in Mathematical Box 5.1. A simple circuit using a transistor is shown in (b). The conventional symbol for the transistor has the emitter, base and collectors arranged as in (a). The input voltage drives a current through a resistor into the base; as a result the supply voltage can drive a current through the device, which generates the output voltage across another resistor. The output voltage is proportional to the input voltage.

power a loudspeaker or similar device. A transistor can also be used as a controlled switch and as such plays an essential part in the operation of computers and other digital devices. In this section we shall describe the construction of a transistor and explain how it can be used in both these ways.

Essentially, a transistor consists of three pieces of doped semiconductor arranged in the series p–n–p or n–p–n. As the modes of operation of these two configurations are essentially the same, we need only discuss one of them. We will choose the first, which is illustrated in Figure 5.7. We refer to the lower p-type region as the 'emitter' because it emits holes and the upper one as the 'collector' because it collects the holes; the central n-type region is known as the 'base'. To operate a transistor, we apply a positive voltage between the emitter and collector. From the earlier discussion of p–n junctions, we see that holes should readily flow across the emitter–base junction, because it is forward biased, but we expect no current to flow between the base and the collector because it is reverse biased. However, an important feature of the design of a transistor is that the base region is deliberately made very thin and is lightly doped with impurities, so that it is possible for at least some of the holes to pass from the emitter, through the base, to the collector without encountering any electrons with which to recombine. As a result, a current can flow round the circuit. Now consider the effect of injecting electrons into the base, which corresponds to drawing a current out of it as in Figure 5.7. Some of the electrons will cross the base and flow through the forward-biased junction between the base and the emitter, while others will be annihilated by combining with holes flowing through the base from the emitter to the collector. If the device is appropriately designed, the resulting current passing through the collector will be very much larger (typically one hundred times) than the base current. Small changes in the base current cause correspondingly large changes in the collector current, so we have a 'current gain' – see Mathematical Box 5.1 This gain depends on the densities of holes in the p-type regions and of electrons in the base region and on the dimensions of the base; provided the base–emitter voltage is not too large, the gain is constant for a given transistor. When larger voltages are applied to the base, the collector current reaches a 'saturation' value and remains at this level even if the base voltage is increased further.

This current gain can be converted to a voltage gain by constructing the circuit shown in Figure 5.7(b). An input voltage drives a current through a resistor into the base of a transistor, which then allows the supply voltage (i.e. a battery or other power source) to drive a current through the transistor from the emitter to the collector and then through a second resistor. An output voltage proportional to the current through this resistor therefore appears across it. As a result, the output voltage is proportional to the input voltage and we have a voltage gain.

MATHEMATICAL BOX 5.1

In Figure 5.7, the base current (I_B) injects electrons into the base region. Some of these pass through the forward-biased junction into the emitter. In any particular semiconductor, the ratio of the number of holes (the 'majority carriers' in this case) to the number of electrons (the 'minority carriers) is fixed, so this part of the base current must be proportional to the emitter current, I_E. Other electrons interact with and annihilate holes that have entered the base from the emitter. The bigger the emitter current, I_E, the more likely is such an interaction, so we can expect this part of the base and emitter currents to be proportional to I_E also. Thus

$$I_B = fI_E$$

where f is a constant. To ensure a large current gain, the system is designed so that f is always a small fraction – i.e. much less than 1. The total current entering the base must be the same as that leaving it; otherwise electric charge would build up, producing an increase in the energy of the system. Hence, if the collector current is I_C, we have

$$I_C = I_E + I_B = (1/f + 1)I_B$$

So that

$$I_C/I_B = (1 + f)/f$$

As f is much less than 1, the current gain is large and approximately equal to $1/f$. In typical devices, f is about 0.01, corresponding to a current gain of around 100.

We now turn to the question of how a transistor can be used as a controlled switch. Referring again to Figure 5.7(b), the principle is simply that we apply either a zero voltage or a large voltage to the input, which results in either a very small current or a comparatively large (i.e. saturation) current flowing through the emitter. Thus the emitter current, and therefore the output voltage, is switched on or off depending on the size of the base voltage. This principle can be applied to construct the basic operational units in a digital computer. We first note that any number can be represented by a series of 'binary bits', each of which can have the value 1 or 0. Following this convention '10' represents $1 \times 2 + 0 \times 1$ (i.e. 2), while '100' represents $1 \times 2 \times 2 + 0 \times 2 + 0 \times 1$ (4) and so on. Thus, for example,

$$101101 = 1 \times 32 + 0 \times 16 + 1 \times 8 + 1 \times 4 + 0 \times 2 + 1 \times 1 = 45$$

Any physical system that can exist in either one of two states can be used to represent a binary bit. Applying this principle in the transistor context, we adopt the convention that a voltage greater than some threshold represents 1, while one less than the threshold represents 0. Thus the numerical value of a binary bit represented by the output voltage in Figure 5.7(b) will be 0 if the input voltage represents 0 and 1 if it corresponds to 1.

As a further example, let us consider one of the basic computer operations: the 'AND' gate, which is a device in which an output bit equals 1 if and only if each of two input bits is also 1. Such a device can be constructed using two transistors as shown in Figure 5.8. A current will flow through both transistors only if both base currents are large enough. For this to happen both input voltages must be large, implying that the digits represented by them are both 1. In this case, the digit represented by the output voltage

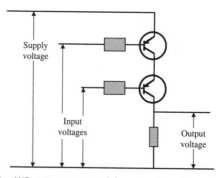

Figure 5.8 An AND gate constructed from two transistors. A significant collector current can flow only if both base currents are sufficiently large, so the output voltage will be small and represent 0 unless both input voltages are large enough to represent 1, in which case the output voltage will also be 1.

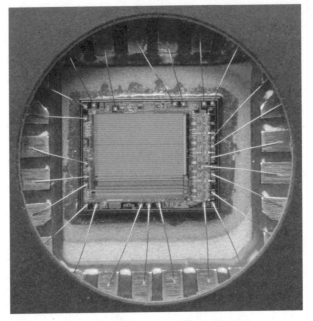

Figure 5.9   A silicon chip used as a computer processor.

will also be 1, but it will be 0 otherwise. This is just the property needed for an AND gate. Similar circuits can be designed to perform the other basic operations such as 'OR', where the output is 1 provided either input is 1, and 0 otherwise. All computer operations, including those used in arithmetic, are built up from combinations of these and other similar basic components.

Computers built soon after the transistor was developed in the 1950s and 1960s were indeed constructed from individual transistors along the lines described above. However, as their operation became more sophisticated, large numbers of transistors were needed to meet their requirements. A major development in the mid 1960s was the invention of the 'integrated circuit', in which many circuit components such as transistors and resistors, as well as the equivalents of the wires connecting them, were contained on a single piece of semiconductor, known as a 'silicon chip' (Figure 5.9). As the technology improved, it became possible to reduce the size of the individual components and so have more of them on each chip. This had the added advantage that the switching times could also be made smaller, so computers have become steadily more powerful and faster over the years. The Pentium 4 processor in the computer on which I am writing this text consists of a silicon chip of about one square centimetre in area; it has around 7.5 million circuit elements, many less than 10^{-7} m in size and the basic switching time or clock speed is around 3 GHz (i.e. 3×10^9 operations per second). However, 10^{-7} m is several hundred times the atomic separation, so each element can still be considered as a crystal and the transistors in a silicon chip operate by the same quantum physics principles as we have discussed in this chapter.

the photovoltaic cell

A photovoltaic cell is a device based on semiconductors which converts the energy from sunlight into electricity. Because all the energy comes from the sunlight that would strike the Earth in any case, it does not contribute to the greenhouse effect and it does not consume any of the Earth's reserves of fossil or nuclear fuel. Various such devices have been developed over the years, the research being driven by the wish to develop this non-polluting form of energy to the point where it can satisfy a significant part of human energy consumption. Photovoltaic cells are all composed of semiconductors. When a photon of the right energy strikes a semiconductor it can cause an electron to be excited to the upper band, leaving a positive hole in the lower band. To produce a voltage, we need the electron and hole to move apart from each other and drive a current through an external circuit. One way to achieve this is to use a p–n junction. As noted above, in the junction region where the p-type and n-type materials meet, there are very few charge carriers because the electrons and holes cancel each other out; and there are excess positive and negative charges on the n-type and p-type sides, respectively, of the interface (see Figure 5.10). If we shine light on this junction area the photons may be absorbed, exciting electrons from the valence to the conduction band and creating an electron–hole pair. These may recombine quickly, but there is a significant

semiconductors and computer chips

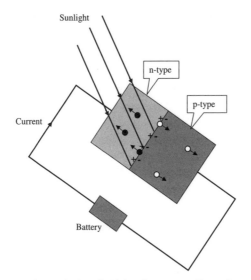

Figure 5.10 A photovoltaic cell. Light photons are directed on to the junction region of a p–n junction, where they excite electrons from the valence to the conduction band, creating an electron–hole pair. Some of these pairs recombine immediately, but some electrons (filled circles) move into the n-type material and some holes (open circles) move into the p-type regions. As a result a current flows, which can be used to charge a battery. Note that the n-type material is in practice made much thinner than shown to allow light to penetrate to the interface region.

probability that instead the electrons will be accelerated into the n-type region by the electrostatic forces acting on them, while the holes are being similarly accelerated into the p-type material. As a result the device can drive a current through an external circuit to charge a battery.

The practical construction of photovoltaic cells involves producing layers of silicon that are so thin that light can pass through them into the junction region. To be useful, they must be as efficient as possible – i.e. a large fraction of the incident photons should create electron–hole pairs – and cheap enough to compete with other forms of energy production. Considerable progress is being made on this at the present time, and photovoltaic cells could well make an important contribution to the generation of green energy by the second decade of the twenty-first century.

summary

In this chapter we have discussed how quantum physics underlies the operation of semiconductors, which in turn underlie much of our modern information and computing technology. The main points are:

- A semiconductor is similar to an insulator, but the energy gap is small enough to allow some electrons to be thermally excited across the gap into the conduction band.
- Both the excited electrons and the positive holes created in the lower band can conduct electricity.
- By adding controlled amounts of appropriate impurities, 'n-type' or 'p-type' semiconductors with an excess of electrons or holes, respectively, can be created.
- A p–n junction formed by joining a p-type to an n-type semiconductor has the properties of a rectifier, which means that it conducts current in one direction only.
- Joining three pieces of semiconductor in the sequence p–n–p or n–p–n makes a transistor, which can act as an amplifier or a switch.
- When used as switches, transistors can be used to represent and manipulate binary bits, which is the basis of electronic computing.
- When light shines on a p–n junction, electron–hole pairs are produced. These can be used to produce an electric current. The resulting device is known as a 'photovoltaic cell'.

chapter six

superconductivity

In Chapter 4 we saw how the fact that electrons can behave as waves allows them to pass through a perfect crystal without bumping into the atoms on the way. Provided, as in a typical metal, there are empty states available, the electrons will respond to an applied field and a current will flow; whereas in an insulator the presence of an energy gap means that there are no such empty states and therefore no current flow. We also saw that in practice, the current flow through a metal experiences some resistance because all real crystals contain imperfections associated with the thermal displacement of atoms from their normal positions in the crystal and the replacement of some of the atoms by impurities. In this chapter we shall discuss another class of substances, known as 'superconductors', in which resistance to current flow completely disappears and electric currents, once started, can flow indefinitely. Ironically, we shall find that this behaviour also results from the presence of an energy gap that has some similarities to, as well as important differences from, the energy gap that prevents current flow in an insulator.

Superconductivity was discovered, more or less by accident, by a Dutch scientist, Kamerlingh Onnes, in 1911. He was conducting a programme of measurement of the electrical resistance of metals at temperatures approaching absolute zero, which had recently become accessible thanks to technological developments in the liquefaction of gases. Helium, in particular, liquefies at only a few degrees above absolute zero at normal pressure and can be cooled even further by reducing its pressure using vacuum pumps. Onnes

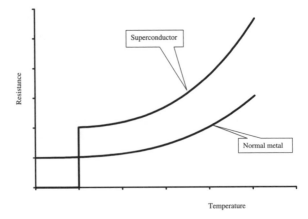

Figure 6.1 The variation of the electrical resistance with temperature for a normal metal such as copper and a superconductor such as lead.

found that the resistance of all metals decreases as the temperature is lowered, but in most cases there is still some resistance to current flow at the lowest temperatures available and, by inference, at absolute zero (this is illustrated in Figure 6.1). We can understand this behaviour in terms of the model we developed in Chapter 4. As the temperature reduces, so does the thermal displacement of the atoms from their average positions, and there is less likelihood of these interacting with the electrons; however, the resistance owing to impurities is unaffected by temperature and is still present at absolute zero.

It was when Onnes focused his attention on the metal lead that surprising results emerged. As shown in Figure 6.1, at comparatively high temperatures lead has a higher resistance than copper, but as the temperature drops below about 4 K (i.e. four degrees Celsius above absolute zero – see Chapter 1) the electrical resistance of lead suddenly disappears. This does not mean that it just becomes very much smaller than that of copper – it is literally zero. This is very unusual in science. Normally 'zero' means much smaller than any comparable quantity in the same way as 'infinite' means something that is much larger than other similar quantities, but in this case 'zero resistance' really does mean what the term implies. As we shall see, this is another consequence of quantum physics impinging on the everyday world.

Although the quantum physics of metals and insulators was worked out in the 1930s – i.e. within about ten years of the discovery of quantum theory itself – it was a further twenty years or so before the phenomenon of superconductivity was properly understood. This was eventually achieved by three scientists: John Bardeen, Leon Cooper and John Schrieffer, who were awarded the Nobel Prize in 1972. This was the second time Bardeen had a share in a Nobel Prize: in 1956 he had shared one with William Shockley for the invention of the transistor. The set of people who have had a share in two Nobel Prizes for physics is very exclusive, as John Bardeen is its only member! The 'BCS theory of superconductivity', as it is now called after the names of its inventors, relies on two main new ideas. The first is that there is a weak attractive force between electrons in a metal, despite the fact that they are all negatively charged and like charges repel. The second is that in a superconductor the presence of this interaction results in electrons being paired up and that the interaction between these pairs leads to the creation of an energy gap that prevents the paired electrons colliding with thermal excitations or impurities, but still allows current to flow. This is in complete contrast to the energy gap in insulators, which, as we saw in Chapter 4, is instrumental in preventing the flow of current.

A word of warning: superconductivity is a subtle phenomenon and a full understanding requires quantum calculations that are considerably more sophisticated than those needed to explain anything we have discussed so far. This means that our treatment will be even less rigorous than usual and we can only expect to get a flavour of the full quantum treatment, and will have to accept quite a lot on trust.

We consider first how it can be that electrons have a net attraction for each other. As discussed in Chapter 2, if two electrons were on their own in free space, the only force between them would be the electrostatic repulsion between them, which is represented by a potential energy that is inversely proportional to the distance separating them – see Figure 6.2(b). However, the electrons we are concerned with are not in free space; they are in a

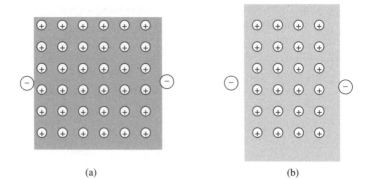

(a)   (b)

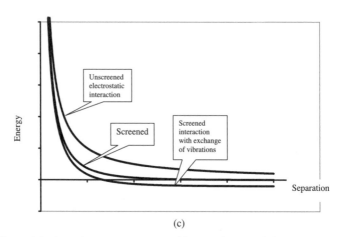

(c)

Figure 6.2   Two electrons are separated by a region containing positively charged ions and a cloud of negatively charged electrons that are represented by the grey tint in (a) and (b). In (b) the electrons come closer together, which pushes out the electron cloud so that it becomes thinner and no longer completely neutralises the ions. The effect of this 'screening' on the effective force between the two electrons is illustrated graphically in (c), which also shows the effect of adding an attractive potential resulting from the exchange of thermal vibrations.

metal along with a whole lot of similar electrons and positively charged atomic nuclei. This changes the situation in several ways. The first effect we shall consider is known as 'screening', which is closely associated with the fact that metal boxes are used to protect or 'screen' sensitive electronic equipment from unwanted electric fields. Referring to Figure 6.2, we focus our attention on two electrons in a metal which are separated by a region of positively charged ions in the metal crystal. Imagine that we try to push the electrons towards each other; this increases the potential of interaction between them, but it also has the effect of applying a force to the remaining electrons, which tend to get pushed out of the space between those we are considering, with the result that they no longer completely cancel out the positive charge in this region. These positive charges then act to attract the electrons we are considering towards the intermediate space and hence towards each other. The net effect of this is that the effective repulsion between the two electrons is greatly reduced from what it would have been at the same separation in empty space. This is the effect known as 'screening' because the response of the intervening material effectively screens the charges from each other. Screening is more effective the greater the separation between the electrons. The net effect is that the effective interaction potential is very much reduced at large separations – see Figure 6.2. However, we should note that even the screened potential is everywhere still repulsive and does not produce the net attraction between the electrons required for superconductivity.[1]

To get some idea of how the effective interaction between two electrons in a metal can be attractive rather than repulsive, we need to consider a much more subtle effect that depends on the electron's quantum wave properties. Consider an electron moving through a crystal lattice. Although it is much lighter than the atoms it passes, it still has a small effect on them and they tend to vibrate a little in sympathy with the electron wave. This does little in itself, but if there are two electrons it is possible for the atomic vibrations set up by one of them to affect the other and vice versa. Detailed quantum calculations show that the effect of this exchange of lattice vibrations between the electron waves is to reduce the total energy very slightly below what it would have been if the waves had not been present and that this attractive potential is largely independent of the electron separation. Provided the electrons are far enough apart for the screened repulsion to be sufficiently small, the overall result can be an attraction, as illustrated in Figure 6.2(c).

The attractive potential is extremely weak and very much less than the typical thermal energies even at the low temperatures where superconductivity occurs. However, the interaction is essentially a quantum one that is best described as being between the electron waves. As these waves are spread throughout the whole crystal, the interaction is essentially independent of the electrons' position, as indicated in Figure 6.2(c). At first sight, this might lead us to conclude that every electron is interacting with every other electron, resulting in a lowering of the total energy. This would therefore have little or no effect on the physical properties of the system. However, one of the results to emerge from a full quantum calculation is that only those electrons whose energies differ from the Fermi energy by less than the energy of a typical lattice vibration (which is about one thousand times smaller than the Fermi energy) interact significantly. When we put all this together and apply quantum physics we find that the attraction is effectively concentrated on pairs of electrons that are moving at the same speed in opposite directions and we shall now try to understand this in a little more detail.

For the reasons just stated, we concentrate only on electrons with energies close to the Fermi energy. All these electrons are moving at about the same speed (and therefore have the same wavelength) although in different directions. This is one point where the one-dimensional model we have used to illustrate the properties of metals and semiconductors breaks down. In one dimension, only two travelling waves have the same wavelength: one travels to the right and the other to the left. However, in three dimensions, waves can travel in all directions, so there is a large number of electrons associated with waves of the same or similar wavelength; this is illustrated in two dimensions in Figure 6.3. Now consider a pair of electrons selected from this set: the two electrons have the same speed but are moving in different directions. The total velocity of a pair is calculated by adding together the individual velocities of the two electrons, remembering to take account of their direction. If we consider any general pair, such as the electrons with velocities v_1 and v_2 shown in Figure 6.3(a), this total velocity is quite different from that of any other pair. However, an exception to this is the special case where the two electrons are moving in diametrically opposite directions, because

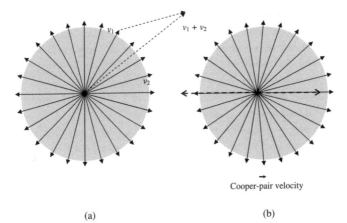

$\overrightarrow{\hspace{1cm}}$
Cooper-pair velocity

(a)   (b)

Figure 6.3 Cooper pairs. In (a) the arrows represent electrons with the
Fermi energy which are moving with the same speed but in different
directions. All states with smaller energy and hence lower speed are
occupied by other electrons and are contained within the shaded circle. No
current is flowing and the average velocity of the electrons is zero. The
dotted lines show how to calculate the total velocity of the pair of
electrons with velocities v_1 and v_2. The total velocities of all such pairs are
different, unless the individual velocities are equal and opposite, in which
case their total velocities equal zero and the electrons are bound into
Cooper pairs. In (b) a current flows from left to right and the velocity of
all electrons is increased by the same small amount (much exaggerated in
the diagram). The total velocity of all Cooper pairs is the same – i.e. twice
the change in velocity of a single electron – and the binding energy is
unchanged from the zero-current case. The thick broken lines, with open
arrowheads illustrate the effect of reversing the velocity of one of the
Cooper pairs: the energy of one electron is increased and that of the other
is reduced. However, the latter state is already occupied, so this process is
forbidden by the Pauli exclusion principle.

here their total velocity is zero in every case. The result that comes
from the more sophisticated quantum analysis is that the effective
binding energy of such a pair equals the very weak interaction
energy described above multiplied by the number of such pairs
that all have the same total velocity. It follows from the above that
the small attractive interaction will be concentrated or amplified
into creating pairs of electrons with zero net velocity, and that

significant energy is required to break such a pair apart into its component electrons. The first person to understand this was Leon Cooper (the 'C' in BCS) and the pairs of coupled electrons are commonly known as 'Cooper pairs'. An additional property of Cooper pairs that comes from the full theory is that the electrons forming such a pair have opposite spin, which means that the total spin of a Cooper pair is zero.

We can now see how the formation of Cooper pairs leads to the formation of an energy gap. This is because the only way that a collection of Cooper pairs can absorb energy is if one of the pairs is broken so that the electrons move independently, and to break such a pair requires a minimum amount of energy. This is reminiscent of the energy gap in insulators discussed in Chapter 4, but differs from that in an important respect, which completely reverses its consequences.

In an insulator, the electron waves interact with the periodic array of atoms and a gap appears when there is a match between the electron wavelength and the atomic separation. In a superconductor the gap comes about because electrons with the same wavelength interact with *each other*. Consider what happens in a superconductor if all the electrons change their velocities by the same (small) amount, as shown in Figure 6.3(b). The total velocity of a Cooper pair is no longer zero, but it is the same for every pair, so the arguments above that led to pairing still apply: the electrons remain paired and the energy gap is unaffected. Moreover, as all the pairs have the same total velocity, a net current is flowing, which, once established, cannot be disrupted by collisions between electrons and obstacles such as impurities or thermal excitations of the crystal lattice. To understand this, consider two ways in which we might think that the current flow would be affected by such collisions: first where the electrons remain paired, and second where a collision leads to pair breaking. In the first case, we might imagine the velocity of a particular Cooper pair being changed by one of the electrons colliding with an obstacle, such as an impurity or a thermal defect in the lattice (cf. the discussion of electrical resistivity in Chapter 4). However, as we see from Figure 6.3(b), when a current flows, the two electrons making up a pair have different speeds, and if the net velocity is reversed, one of the electrons will have to end up in a state that is already occupied – and this is forbidden by the exclusion principle.[2]

Another mechanism that might be thought capable of causing resistance is one in which a Cooper pair is broken into its component electrons, which then collide with impurities or thermal defects as in a normal metal. However, in order to break a Cooper pair, an amount of energy at least as great as the size of the energy gap must be supplied. At any finite temperature, this may take the form of thermal energy and there is always some probability of a thermal excitation leading to pair breaking. However, the electrons soon combine to reform the Cooper pair so at any one time only a fraction of the Cooper pairs are split, and the remainder continue to transport the current without resistance. As the temperature increases, the fraction of broken pairs becomes larger and this trend is reinforced by the fact that the size of the energy gap reduces. At a particular 'critical' temperature, the energy gap falls to zero: at this point and above, Cooper pairs can no longer exist and the material behaves like a normal metal. This is illustrated in the case of lead in Figure 6.1.

We conclude therefore that a superconductor can support the flow of electric current without any resistance because of the formation of an energy gap that persists in the presence of a current. Thus, once a current is established, it flows essentially for ever unless the superconductivity is destroyed – for example, by raising the temperature. This property of a superconductor has been tested experimentally by measuring the current flowing round a superconducting loop as a function of time. No detectable change was found and the sensitivity of the measurement was such that it was calculated that no significant decay would be expected if the experiment continued for hundreds of years. This differs greatly from the behaviour of metals such as copper, where even in the case of very pure samples at very low temperature, currents decay in a small fraction of a second when the driving voltage is removed.

So how are currents started and stopped in superconductors? Consider the electrical circuit illustrated in Figure 6.4. In Figure 6.4(a) a battery or other power source drives a current round a

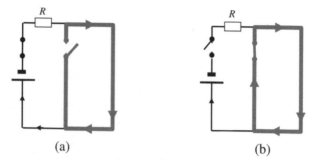

(a)   (b)

Figure 6.4 Current flow in a superconducting circuit. Black lines represent normal conductors and the thick grey line represents a superconductor. In (a) the voltage produced by the battery drives a current through the resistor, R, and round part of the superconducting circuit. In (b) one switch is closed to allow current to flow round the complete superconducting circuit, and the other is opened to detach the battery. Current continues to flow indefinitely as long as this arrangement is maintained.

circuit, part of which is made from superconducting material. A switch is now closed (Figure 6.4[b]) to complete a circuit and another switch is opened to disconnect the battery. Current now flows round the superconducting circuit and continues flowing. An alternative method is to use the fact that a current can be induced in a loop of wire by subjecting it to a changing magnetic field. This can be done by moving a magnet through a loop of wire: if the wire is a normal conductor, current will flow round the loop as long as the magnet is moving and causing the field to change, but as soon as the motion stops, the current stops flowing because of the wire's resistance. However, a superconductor has no electrical resistance, so the current keeps flowing after the magnet has stopped moving and ceases only if the magnet is returned to its original position or if the superconductivity is destroyed.

This property of superconductors is dramatically illustrated by the phenomenon known as 'magnetic levitation' or the 'floating magnet'. If a piece of superconductor is lying horizontally and we bring a magnet down towards it, currents are induced which in

Figure 6.5 Magnetic levitation: a magnet floats above a piece of high-temperature superconductor. (Photographed in the University of Birmingham, with assistance from Chris Muirhead).

turn produce a magnetic field that exerts a force on the magnet. By a principle known as 'Lenz's law', this force opposes the magnet's original motion and therefore counters the force of gravity pulling the magnet down. The more powerful the magnet, the larger is the current and the greater is the vertical force. If the magnet is powerful enough and light enough, this force can balance the gravitational force so that the magnet is held above the surface of the superconductor – i.e. is levitated, as illustrated in Figure 6.5.[3] Using modern superconductors and magnets, which were both developed towards the end of the twentieth century, these levitation forces can be large enough to support quite heavy weights.

Because supercurrents flow without resistance, they have potential applications in any situation where a substantial electric current is used. An obvious example is the distribution of electricity from power station to consumer. Because conventional conductors have resistance, a significant amount of the distributed power (typically about thirty watts per metre of cable) produces heat in the wires, which is dissipated into the air. If superconductors could be used, most or all of this wastage could be avoided.

However, there is one main obstacle to doing this in practice: superconductivity occurs only at very low temperatures, so the energy and cost of the refrigeration needed to bring the wires into a superconducting state and keep them there is normally greater than any saving achieved.

A well-established major application of superconductivity is to the construction of large magnets. Because supercurrents flow without resistance, the magnetic fields produced by them require no power to sustain them once they are established. Hence, no heat is produced in the magnet windings and this allows fields to be generated that are so large that the associated currents would melt coils produced from conventional materials such as copper. In contrast to power lines, which may extend for hundreds of miles, magnets occupy a comparatively small space – rarely larger than a room in a typical home – so refrigeration is comparatively economic. For all these reasons, superconducting magnets are now commonly used in situations where large, stable magnetic fields are required. An example of this is the use of superconducting coils in the magnets used in the nuclear magnetic resonance scanners used in medicine.

Magnetic fields also play an important role in the operation of electric motors, and in principle, superconductors could play a useful role here. However, this is likely to be a quite specialist application, since typical thermal losses in a conventional motor are small compared with the total power of the motor, and the magnetic fields produced are usually not as large as those in the magnets discussed above. The main advantage of a superconducting motor would be its small size, which results from the fact that superconducting wires are generally much smaller than the equivalent copper conductors.

'high-temperature' superconductivity

We mentioned above that superconductivity was first discovered in lead when Kamerlingh Onnes used liquid helium to cool the material down to below 4 K. In the seventy-five years that followed this discovery, superconductivity was discovered in a number of metals and alloys, but the highest critical temperature was less than 23 K, and achieving this temperature still required the use of liquid helium. Even today, helium is a gas that is difficult and expensive to liquefy. To prevent the liquid boiling, it must be surrounded by two vacuum flasks with the space between them filled with liquid nitrogen. Thus, until the 1980s superconductivity was regarded as a pure-science topic with only the most specialist applications. Then, in 1986, 'high-temperature' superconductivity came along.

J. Georg Bednorz and Karl Alex Müller were two scientists who worked for IBM in Zurich. They recognized the potential of a superconductor that would operate at temperatures higher than that of liquid helium and, almost as a spare-time activity, they began a programme of testing different materials to see if any would live up to this dream. They were probably as surprised as anyone else when having turned their attention to a particular compound of lanthanum, bismuth, copper and oxygen, they found that its electrical conductivity dropped sharply to zero when it was cooled below 35 K – which, although still a very low temperature, is more than one and a half times the previous record.

This pioneering work was quickly built on by others and in January 1987 a research team at the University of Alabama—Huntsville substituted yttrium for lanthanum in a compound similar to that discovered by Bednorz and Müller and found that the compound was superconducting up to 92 K. Not only was this another major advance up the temperature scale, it also passed an important milestone – the boiling point of nitrogen, which is 77 K. This meant that superconductivity could now be demonstrated without the use of liquid helium. Liquid nitrogen is much easier to produce than liquid helium, is more than ten times cheaper and can be stored and used in a simple vacuum flask. For the first time, superconductivity could be studied without expensive, specialist equipment; superconducting phenomena such as magnetic levitation that had previously been observed only through several layers of glass, liquid nitrogen and liquid helium could be seen on the laboratory bench (an example of this appears in Figure 6.5). Progress since 1987 has been less dramatic. The highest known transition to the superconducting state occurs for a compound of the elements mercury, thallium, barium, calcium, copper and oxygen at 138 K at normal pressures; under extreme pressure its transition temperature can be raised further – to over 160 K at a pressure of 300,000 atmospheres.

The fact that the transition temperatures of these compounds are so much higher than those previously observed has led to them being called 'high-temperature superconductors'. This title is potentially misleading, for it seems to imply that superconductivity should occur at room temperature or even higher, which is certainly not the case. However, the highest superconducting temperature was raised from 23 K to 92 K – i.e. by four times – between 1986 and 1987; if a further factor of three could be achieved, the dream of a room-temperature superconductor would have been achieved. We might have expected that the advance to liquid nitrogen temperatures would have greatly increased the potential for practical applications of superconductivity, but these have been less dramatic than was originally hoped. There are two main reasons for this. First, the materials that constitute high-temperature superconductors are what are known as 'ceramics'. This means that they are mechanically similar to other ceramics (such as those found in kitchens) in that they are hard and brittle and therefore very difficult to manufacture in a form suitable to replace metal wires. The second problem is that the maximum current that a high-temperature superconductor can support is rather too small for it to be of practical use in the transport of electricity or the production of large magnetic fields. However, this is still an area of active research and development. The design of motors based on high-temperature superconductors, for example, has reached the prototype stage in the early years of the twenty-first century. Their greatest potential is where high power combined with low weight is required: for example an electric motor to power a boat.

flux quantization and the Josephson effect

We have seen that superconductors contain Cooper pairs in which the electrons are bound together. As a result, the quantum physics of superconductors can be conveniently described as the motion of such pairs rather than of the individual electrons. Such a pair can actually be thought of as a particle with a mass equal to twice the electron mass and charge equal to twice the electron charge, moving at a velocity equal to the net velocity of the pair. The wavelength of the matter wave associated with such a particle can be calculated from the velocity and mass of the pair using the de Broglie relation (cf. Mathematical Box 2.3). The fact that current can be transported through a superconductor by Cooper pairs that are not scattered by obstacles means that the quantum wave representing them extends 'coherently' through the whole crystal. This is in contrast to a normal metal, where the wave essentially breaks into pieces every time an electron is scattered by a thermal defect or an impurity, as discussed in Chapter 4. One result of this coherence is what is known as 'flux quantization'. To understand this, we first have to know a little more about magnetic fields. Referring to Figure 6.6, we see that when an electric current flows around a loop of wire it generates a magnetic field (labelled B) through the loop. The total field added up over the whole area of

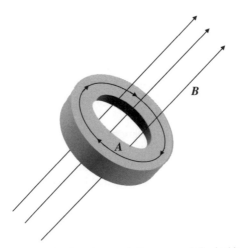

Figure 6.6 Flux quantization. A current flows around the inside surface of a superconducting ring, creating a magnetic field, B, threading the ring. Associated with B is the vector potential A, which has the form of closed loops inside the superconductor. The total of A summed around one of these loops equals the total flux of B through the ring.

the loop is known as the 'magnetic flux'. Suppose now that the loop of wire is in fact a superconductor. The wave function representing the Cooper pairs must join up on itself so that the distance round the loop equals a whole number of wavelengths (cf. Figure 4.2). For rather subtle reasons, outlined in Mathematical Box 6.1, this places constraints on the value possessed by the magnetic field through the loop: its flux always equals a whole number times the 'flux quantum', which is defined as Planck's constant divided by the charge on a Cooper pair. This works out as equivalent to a field of magnitude about two millionths of the Earth's magnetic field passing through an area of one square centimetre.

We turn now to what is known as the 'DC Josephson effect', which involves the tunnelling of Cooper pairs. We discussed quantum tunnelling in Chapter 2, where we saw that wave-particle duality means that particles such as electrons can penetrate a potential barrier where this would be impossible classically. Such a barrier can be created if two pieces of superconductor are brought close together but separated by a narrow piece of insulating material. Provided the barrier is narrow enough and the current is less than a maximum value (known as the 'critical current'), Cooper pairs can tunnel through the barrier while preserving their identity and coherence. This idea greatly surprised the superconducting community when Brian Josephson predicted it theoretically in 1962, but since then it has been firmly established experimentally and fully understood as another manifestation of quantum physics.

There are two main applications of the Josephson effect, neither of which is susceptible to a simple explanation, so we shall have to be content with a description only. The first relates to the accurate measurement of magnetic field by what is known as a 'superconducting quantum interference device' (SQUID). This consists of a superconducting loop interrupted by two Josephson junctions, as in Figure 6.7. A current passes from one side of the loop to the other, being divided into two parts as it does so, one of which passes through each junction. As the currents have wave properties, there can be interference between the two waves and the net result is that the maximum current that can pass through the circuit without destroying the superconductivity depends on

MATHEMATICAL BOX 6.1

To get some understanding of flux quantization, we first define a further quantity, A, known as the 'vector potential' associated with the magnetic field. If we consider any loop that has a magnetic field passing through it, then A points along this loop as in Figure 6.6 and its magnitude is such that the magnetic flux in the loop (f) equals A times the length of the loop (L). Thus

$$f = AL$$

For a wave to fit coherently round a ring of circumference L, this distance must equal a whole number of wavelengths (l), so that

$$L = nl = nh/p$$

where n is an integer, p is the Cooper pair momentum and the last result comes from the de Broglie relation discussed in Chapter 2

To understand how this leads to flux quantization, we have to know that the momentum of a charged particle moving in a magnetic field is not just the normal expression mv, but contains an additional component equal to qA, where q is the charge on the particle. Also, when a current flows in a superconductor, it does so only near the surface of the material, so qA equals the entire momentum in the case of a Cooper pair in the body of the superconductor. Putting all this together and referring to Figure 6.6, we get

$$L = nh/qA$$

so that

$$f = AL = nh/q = n(h/2e)$$

using the fact that the magnitude of the charge on a Cooper pair is twice that on a single electron (e). The quantity $h/2e = 2 \times 10^{-15}\,\mathrm{J\,s\,C^{-1}}$ is known as the 'flux quantum' and the above shows that the flux through a loop of superconductor always equals a whole number of flux quanta. The magnitude of the Earth's magnetic field at a point on the surface is about 5×10^{-5} of the same units per square metre, which is equivalent to 4×10^5 flux quanta per square centimetre.

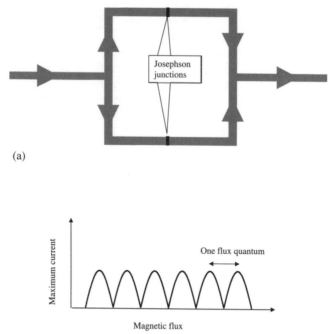

(a)

(b)

Figure 6.7 A SQUID. In (a), a current flowing through a superconductor (represented by the thick grey lines) splits into two parts, each of which passes through a Josephson tunnel junction. When the waves associated with the currents are reunited, the quantum interference between them results in the maximum allowed value of the current oscillating when the magnetic flux through the ring changes (b).

the amount of magnetic flux penetrating the loop. This maximum current oscillates as the magnetic field changes. The period of this oscillation is one flux quantum (see Figure 6.7[b]). If we place the SQUID in a region penetrated by a magnetic field, it is then possible to measure the size of the flux through the loop with an error considerably less than one flux quantum. For a typical field used in a laboratory, this is an accuracy of better than one part in 10^{10}, which far exceeds that of any other technique for measuring magnetic field.

The second application is known as the 'AC Josephson effect'. As they encounter no resistance, currents can flow through superconductors and Josephson junctions without needing any voltage to drive them. However, the AC Josephson effect occurs when we deliberately apply a constant voltage across a Josephson junction. We then find that the resulting current is not steady but oscillates in time at a frequency equal to the applied voltage multiplied by twice the electronic charge and divided by Planck's constant. For a voltage of ten microvolts, this corresponds to 4.8 billion oscillations per second, which is similar to the frequency of electromagnetic radiation in the microwave range. Such frequencies can be measured with extremely high accuracy, so this effect can be combined with the known values of the fundamental constants to produce an extremely accurate measure of the voltage: so accurate, in fact, that the fundamental standard of voltage is now defined internationally in terms of the Josephson effect.

SQUIDS and other Josephson devices were first developed before high-temperature superconductors were discovered, but this is one area where these materials have been successfully applied to the point where they are produced and sold commercially.

summary

In this chapter we have discussed superconductivity and its applications. The main points are:

- When some materials are cooled to low temperature, they suddenly lose all resistance to the flow of electric current. This phenomenon is termed 'superconductivity' and materials displaying these properties are known as 'superconductors'.
- The electrostatic repulsion between electrons in a metal is reduced at large distances because it is screened by the response of the other electrons and ions between them.
- An electron moving through a metal may set up small vibrations in the lattice of ions; these may interact with another electron, producing an effective attraction that can be greater than the screened repulsion.

- This attraction results in the formation of Cooper pairs, which are composed of two electrons that move in opposite directions with the same speed. Most of these pairs remain unbroken when a current flows, which results in superconductivity.
- Applications of superconductivity include their use in the construction of large magnets, such as those used in magnetic resonance imaging (MRI) scanners.
- High-temperature superconductors operate at temperatures up to about 100 K, in contrast to conventional superconductors which lose their superconductivity above about 20 K or less.
- When current flows round a superconducting loop, the total magnetic flux through the loop is quantized in units of Planck's constant divided by twice the electronic charge.
- Flux quantization leads to the Josephson effect, which is used in SQUIDs to make very precise measurements of magnetic fields.
- The AC Josephson effect can be used to make voltage measurements that are so precise they are used to define the voltage standard.

notes

1. One of the simplifications underlying the above account of screening is the distinction between the two electrons whose interaction we are considering and the others that move away to produce the screening effect. In fact, screening is a dynamic process in which all the electrons are both screening and being screened continuously. A full mathematical analysis of this process shows that it leads to an effective interaction potential of the form illustrated in Figure 6.2(b).

2. In principle, resistance to current flow could occur if all the pairs suffered collisions that changed all their velocities by the same amount at the same time. However, as there are around 10^{20} Cooper pairs in a typical superconducting sample, the probability of this occurring is vanishingly small.

3. An important difference between the magnetic levitation of superconductors and other methods using magnetic forces is

that the superconducting case is stable. Imagine, for instance, trying to balance one magnet on top of another by bringing the two north poles together: the upper magnet will always try to turn round and bring its south pole nearer the north pole of the upper magnet. Most non-superconducting levitation arrangements (e.g. Maglev trains) contain devices that continuously detect and counteract such instability.

spin doctoring

During the last decade of the twentieth century and since, there has been increasing interest in the application of quantum physics to the processing of information – in computers for example. We saw in Chapter 5 that modern computers are based on semiconductors, which in turn are governed by the laws of quantum physics. Despite this, these computers are still commonly referred to as 'classical', because, although quantum physics underlies their operation, the calculations are performed in a perfectly classical manner. To understand this further, we must first recall that all information in a conventional computer is represented by a series of binary 'bits' that can equal either 1 or 0. How these are represented is irrelevant to the way they are manipulated to perform calculations. In quantum information processing, however, quantum physics is essential to the actual computing operations: information is represented by quantum objects known as 'qubits', where behaviour is governed by quantum laws. A qubit is a quantum system that can be in one of two states (like a classical bit) and these can represent 1 and 0, but a qubit can also be in what is called a 'quantum superposition' of these states, in which it is, in some sense, both 1 and 0 simultaneously. What this means should become clearer shortly when we consider some specific examples where we shall see that the quantum processing of information can do some things that are impossible classically.

Although there are a number of different quantum systems that could be used as qubits, we shall confine our discussion to the example of electron spin. In earlier chapters we discovered that electrons, and indeed other fundamental particles, have a quantum property that we referred to as 'spin'. By this we mean that a particle behaves as if it were spinning about an axis in a manner reminiscent of the Earth's rotation or that of a spinning top. As so often happens in quantum physics, this classical model is best thought of as an analogy and difficulties arise if we attempt to take it too literally. The important things to note for our purposes is that spin defines a direction in space, which is the axis the particle is 'spinning' around, and that when we measure the spin of a fundamental particle, such as an electron, we find that it always has the same magnitude, while its direction is either parallel or anti-parallel to the axis of rotation. As a shorthand we can say that the spin is pointing either 'up' or 'down';[1] and we saw in Chapter 2 that these two possibilities play an important role in determining the number of particles allowed by the exclusion principle to occupy any given energy state. We see therefore that spin has at least one of the properties required of a qubit: it can exist in one of two states, which can be used to represent the binary digits 1 and 0. We shall now try to understand how it can also be placed in a superposition state and what this means.

What, we may ask, do we mean by 'up' and 'down'? Surely the electron cannot be affected by such a notion, which depends on our experience of living on the Earth's surface and, in any case, the directions we think of as 'up' and 'down' change as the Earth rotates. Why should we not be able to measure spin relative to, for example, a horizontal axis, so that it is either 'left' or 'right'? The answer to this question is that we can measure spin relative to any direction we like, but once we choose such a direction, we always find that the spin is either parallel or anti-parallel to it. However, the act of making such a measurement, destroys any information we may previously have had about its spin relative to some other direction. That is, the measurement appears to force the particle to reorient its spin so as to be oriented either parallel or anti-parallel to the new axis.

How, in practice, do we measure spin? The most direct way is to use the fact that any particle that possesses spin also has an associated magnetic moment. By this we mean that a fundamental particle like an electron behaves like a tiny magnet pointing along the spin axis. Thus, if we can measure the direction of this magnetic moment the result also tells us the spin direction. One way to measure this magnetic moment is to place the particle in a magnetic field that we have generated in the laboratory; if this field gets larger as we move in, say, an upward direction then a magnet pointing in this direction will move upwards, while one pointing down will move downwards. Moreover, the size of the force causing this motion is proportional to the magnitude of the magnetic moment and hence of the spin, which can therefore be deduced from the amount the particle is deflected. This procedure was first carried out in 1922 by Otto Stern and Walther Gerlach, two physicists working in Frankfurt, Germany. They passed a beam of particles[2] through a specially designed magnet which split the particles into two beams, one corresponding to spin up and one to spin down. This is shown schematically in Figure 7.1(a).

As pointed out above, we are free to choose along which direction we make the measurement. We can do this by rotating the magnet so that its measuring direction is, say, from right to left: again, two beams emerge, but now in the horizontal plane though with the same size of separation as in the vertical case. We conclude that the size of the spin is the same as when the measuring direction was vertical, but that it now points to the left or the right. Thus, in a single experiment, we can measure the spin relative to one direction only and we always get one of two results: the spin is either parallel or anti-parallel to the direction we have chosen. Once we have made such a measurement and found that a particular particle's spin is pointing in a particular direction (say, up), if we repeat the measurement we get the same result, so we can reasonably conclude that this particular particle's spin is really up (see Figure 7.1[b]). Similarly, if we re-measure the horizontal spin of a particle that was previously found to be pointing in, say, the right-hand direction, it will again be found to be pointing in that direction and we can attribute this property to it.

Why, therefore, can we not measure both the horizontal and vertical spins of a particle? Surely, all we have to do is to measure one component using a vertically oriented magnet and then pass it through a horizontal magnet to measure its horizontal spin. The problem is that making the second measurement destroys the information we gained from the first one. This is illustrated in Figure 7.1(c), where we consider passing the up-spin particles

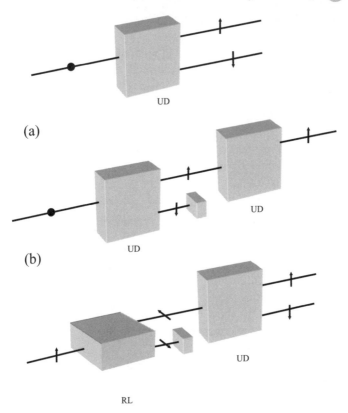

(a)

(b)

(c)

Figure 7.1 The boxes represent Stern–Gerlach magnets. A particle passing through them emerges in one channel or the other depending on the direction of its spin. (a) represents a measurement of the up/down (UD) spin. Once the spin direction is known, subsequent measurements in the same direction give the same answer (b). In (c) particles previously found to have their spins pointing up pass through a left/right (LR) polarizer and emerge from either channel at random. The act of measuring the LR polarization has also destroyed our previous knowledge of its UD state, so the photons again emerge at random when this spin is measured.

emerging from a vertical measurement through a horizontal magnet. We find that half the particles come out through the right-hand channel and the other half through the left, so we are tempted to conclude that we have sorted the particles into two sets, one of which has spin 'up and right' and the other 'up and left'. The reason we cannot do this emerges when we try to confirm the value of the vertical spin by passing, say, the left-pointing particles through a second vertically oriented magnet: instead of all the particles coming through the up channel, half of them come through up and half through down. We are forced to conclude that the act of measuring the horizontal component has destroyed the knowledge we previously had of the vertical spin. As we indicated above, we are up against a fundamental principle of quantum physics: the act of measuring one physical quantity destroys the knowledge we previously had about another. Another way of looking at this is that, before the horizontal spin is measured, a particle with, say, up spin is not 'either left or right', but 'both left and right'. This is what we mean by 'superposition': an up-spin particle is in a state that superposes a left state and a right state; the act of measuring the horizontal spin induces the system to collapse into one or other of the left/right states. We shall return to this point in a little more detail when we discuss quantum computing below.

Examples like the above have consequences for our conceptual understanding of the principles of quantum physics, but we shall postpone consideration of these to the next chapter. For now we shall concentrate on explaining some practical applications of these principles to information processing which have been developed in the late twentieth and early twenty-first century. The two examples we shall concentrate on are 'quantum cryptography' and 'quantum computing'.

quantum cryptography

Cryptography is the science (or art?) of coding messages using a key or cipher so that they can be sent from one person (the 'sender', traditionally named 'Alice') to another (the 'receiver', called 'Bob') while remaining incomprehensible to an 'eavesdropper' ('Eve'). There are a number of ways of doing this, but we shall concentrate on one or two simple examples that illustrate the principles involved and the contribution that quantum physics can make. Suppose the message we want to send is the word 'QUANTUM'. A simple code is just to replace each letter by the one following it in the alphabet – unless it is Z which is 'wrapped round' to become A. More generally, we can encode any message by replacing each letter by the one that is n letters later in the alphabet and wrapping round the last n letters of the alphabet to be replaced by the first n. Thus, we would have

Plain message	Q	U	A	N	T	U	M
Coded using $n = 1$	R	V	B	O	U	V	N
Coded using $n = 7$	X	B	H	U	A	B	T
Coded using $n = 15$	F	J	P	C	I	J	B

This code is of course very easy to crack. There are only twenty-six different possible values of n and it would take only a few minutes with a pencil and paper to try them all; a computer could do this in a tiny fraction of a second. The correct value of n would be identified as the only one to yield a sensible message; the chances of there being more than one of these are very small if the original message is reasonably long (i.e. a few words or more).

A simple but slightly more sophisticated procedure depends on the use of arithmetic. We first replace each letter in the message by a number so that A becomes 01, B becomes 02 and so on, so that Z is represented by 26. We then add a known 'code number' to the message, which can be generated by repeating a shorter number (known as the 'key' to the code) as many times as needed to generate a number as long as the message. This number is written underneath the message and the two rows of digits are added to create the coded message. This procedure is illustrated in the example below, where we have chosen the key to be 537.

Plain message	Q		U		A		N		T		U		M	
As digits	1	7	2	1	0	1	1	4	2	0	2	1	1	3
Code number	5	3	7	5	3	7	5	3	7	5	3	7	5	3
Coded message	7	0	9	6	3	8	6	7	9	5	5	8	6	6

Alice sends the last line to Bob and, provided he knows the procedure and the values of the three digits, he can recover the message by regenerating the code number and subtracting it from the coded message. If Eve intercepts the message and tries to decode it, she will have to try all the one thousand possible values of the key until she sees a meaningful message. Of course, a computer can still do this extremely quickly.

These examples have an important feature in common, which is that the key to the code is much shorter than the message itself. There are much more complex mathematical procedures that can be used. Using a key that consists of about forty decimal digits, a message can be encoded in such a way that a present-day classical computer would have to run for many years to be sure of cracking the code. Thus, if Alice and Bob could exchange a short message in complete secrecy, they could use this to decide which key to use before the message was sent, and then exchange coded messages openly, confident that these will not be understood by Eve. However, this does depend on Alice and Bob knowing the key and Eve having no access to it. It is this secure key exchange that is facilitated by the use of quantum techniques, as we shall now see.

We first remember from Chapter 5 that any number can be written in 'binary' form by a string of 'bits' that have the value 0 or 1. Thus 537 equals $5 \times 10 \times 10 + 3 \times 10 + 7$, but it also equals $2^9 + 2^5 + 2^2 + 1$, which is 1000100101 in binary notation. A typical forty-decimal-digit key consists of about 150 bits when expressed in binary form. To achieve key exchange, all that is needed is that Alice and Bob should both know the value of some 150-bit number, and we should note that neither of them needs to know in advance what the key is going to be. Suppose Alice represents the key by a set of particle spins, assuming that one orientation (say, up) represents 0, while the opposite represents 1. She passes a set of spins through a Stern–Gerlach magnet one at a time, records which channel each emerges from and sends them on to Bob. Provided Bob knows how Alice's magnet was oriented, he can deduce the same spin values by passing the particles through an identically oriented magnet and recording through which channel they emerge. Now suppose that Eve intercepts the particles on their way from Alice to Bob: if she knows how Alice and Bob's magnets are aligned, she can align hers the same way; so whenever she receives a particle she can measure its spin direction and send it on to Bob. In this way Eve can record the message without either Alice or Bob knowing she has done so. However, if she does not know how Alice and Bob's magnets are set up, she can only make a guess, and if she guesses wrong, say by setting her apparatus to read right/left when Alice's and Bob's are set to measure up/down, then she will obtain no information about the up/down value of the spin and hence no knowledge of Alice's message. Moreover, the particles she sends to Bob are either left or right and will in either case emerge randomly from the up/down channels of Bob's magnet. We note that this procedure is essentially a quantum one, because it is the act of Eve making a measurement that disrupts the information going to Bob; if the message were encoded classically, Eve would be able to read it and send it on unaltered. In the quantum case, as Eve does not know the orientation of Alice's magnet, she can obtain no information about the message sent by Alice, although she does disrupt it so that Bob does not receive it either. As a result, when Alice uses the key she believes she has agreed with Bob to encode a message, Bob will try to use the version that Eve has forwarded: this will be useless and he will be unable to read the message. Alice and Bob will then quickly realize that there must be an eavesdropper on the line.

However, as described so far, this method of key exchange is not fully secure. If Alice and Bob always use the same magnet orientation to prepare and measure the spins, then Eve can experiment with different orientations of her apparatus until she finds one that corresponds to theirs. To counter this, Alice and Bob might choose randomly between an up/down and a right/left orientation each time they send a spin, but it would seem that, for this to work, Bob would have to know the orientation of Alice's apparatus so that his could be similarly arranged. Moreover, if they tell each other how their magnets are set up, they would have to do this by sending a message; this could then be intercepted by Eve, who could ensure that her magnet had the same orientation and then be in a position to successfully eavesdrop. There is a way for Alice and Bob to get round this. First, Alice sends Bob a particle, having recorded both its spin direction and the orientation of the magnet used (up/down or right/left), and keeps a record of this information. Then she makes a random choice either to change her magnet orientation or to leave it alone, sends another particle and continues this process. For his part, Bob records the spin of the arriving particles using a magnet in a known orientation, which he also either leaves alone or changes at random between measurements. Statistically, in about half the cases Alice's and Bob's magnets will be oriented the same way, though they will be at right angles to each other in the other half. After the exchange is complete, Alice and Bob communicate publicly and tell each other how their magnets were oriented for each measurement, but not what the measurement result was. They identify a subset of measurements where their apparatuses were oriented the same way and discard the rest. Provided the number of particles transmitted is large enough (somewhat over twice the number of bits needed to encode the key), they will then have a key whose value they both know and which can be safely used to encode their communications. If Eve has been on the line, the key exchange will have been disrupted because she does not know how either apparatus is oriented, so Bob will not be able to read the message, but neither will Eve.

Quantum key exchange based on the principles discussed above was practically demonstrated towards the end of the twentieth century. However, although the principles are as discussed, there are considerable differences in its practical implementation. First, the quantum objects used are not normally electrons or atoms, but photons and the quantity measured is not spin (though it is related to it) but photon polarization, which will be discussed in Chapter 8. A practical problem is associated with the noise associated with the loss of particles in the transmission and the appearance of stray particles from the environment. Additional measurements have to be made to overcome these and as a result typically ten times as many bits must be exchanged before a 128-bit key is reliably transmitted. However, the whole process can be automated and completed in a very short time, and, in this way, quantum key exchange has now been performed over a distance of about fifty kilometres at a rate approaching 10^5 bits per second.

quantum computers

Another example of quantum information processing is the quantum computer. We should note that it is actually wrong to talk about quantum computers in the present tense, since the only such devices that have been built to date are capable of only the most trivial calculations, which would be more easily carried out on a pocket calculator or even by mental arithmetic. Nevertheless, if the technical obstacles could be overcome, quantum computers would have the potential to perform certain calculations very much faster than any conceivable conventional machine. For this reason the prospect of quantum computing has become something of a holy grail in recent years, and large amounts of scientific and industrial investment are being devoted to its development. Whether or not this will pay off remains to be seen.

So how is it even in principle possible to exploit the concepts of quantum physics towards this end? A detailed discussion of this is far beyond the limits of this book, but we can hope to understand some of the basic principles involved. The first essential point is that in a quantum computer a binary bit is not represented by an electric current flowing through a transistor, but by a single quantum object such as a spinning particle – we saw an example of this in the previous section when we discussed quantum cryptography. As before, we will assume that 0 is represented by a particle with positive spin in the vertical direction (spin up) while 1 is represented by a negative spin component (spin down). When a quantum object is used to represent a binary bit in this way, it is commonly referred to as a 'qubit'.

As a first example, we consider how we might perform the 'NOT' operation, which is one of the basic Boolean operations that make up a computation and consists of replacing 1 by 0 and 0 by 1. Remember that a spinning particle behaves like a small magnet. This means that if it is placed in a magnetic field it will want to turn like a compass needle to line up with the field direction. This motion will be resisted by the inertia of the spin,[3] but by applying a carefully controlled magnetic field to a spinning particle, it is possible to rotate the spin through any known angle. If, for example, this angle is 180°, an up spin will be rotated to point down and a down spin will be rotated into the up position, which is just what we need to represent the operation NOT. It can also be shown that all the operations a conventional computer performs on bits can be performed on qubits by subjecting spinning particles to appropriately designed magnetic fields. Some of these involve interactions between the qubits, which is one of the challenges to the practical realization of a quantum computer.

So far, all we have done is explain how a quantum computer based on qubits can do the same thing as a conventional computer does with bits. If this were all there were to it, there would be no reason to use a quantum computer: indeed, there is every reason to believe that a quantum computer used in this way would be very much slower and less efficient than its classical rival. To appreciate the potential advantages of quantum computing, we first have to understand a little bit more about the concept of superposition.

We saw earlier that if we know value of one spin component (e.g. the spin axis is pointing to the right), then if we measure a component in another direction (e.g. up/down) then this changes the state of the particle in such a way that we lose the information we had previously. This means that if a particle is in, say, an up state, it is not meaningful to say whether it is pointing to the left or the right. In many ways this is not surprising: if I have an arrow pointing from left to right, there is no answer to the question 'Is your arrow pointing up or down?' However, when the quantum properties of spinning particles are studied in more detail, it turns out that rather than an up spin being *neither* left *nor* right, it is more correctly described as *both* left *and* right. The quantum state of the up-spinning particle is an addition or superposition of the quantum states corresponding to equal parts of the states corresponding to left and right spin. If we consider a spin whose axis is in neither the vertical nor horizontal directions, but somewhere in between, then this can also be thought of as a superposition of up and down. If the particle is in, for example, a right state, the contributions from the up and down states are equal, but if the spin axis is close to the vertical and the spin state is positive, the superposition will consist of a lot of up and a little down. This means that a qubit can be in a state corresponding to a superposition of the states corresponding to 1 and to 0 and the power of a quantum computer lies in the fact that performing a computation on such a state produces a result that is a superposition of the results of performing separate calculations using the different inputs. Returning to the NOT operation, we see that if we start with a left-pointing spin and rotate it to point to the right, we simultaneously reversed both the up- and the down-spin components of the superposition. We have therefore turned 1 into 0 and 0 into 1 in a single operation: we have performed two calculations at the same time. We can do even better than this if we can extend the superposition principle to states consisting of more than one qubit.

We can illustrate this by considering a very simple program. The input to this program is any of the numbers 0 to 3, represented by three binary bits; this is multiplied by 2 and then output as a three-bit number. Since each qubit has two states corresponding to 0 and 1, there are four different calculations we can do:

Input number	Input bits	Input qubits	Output qubits	Output bits	Output number
0	000	↑↑↑	↑↑↑	000	0
1	001	↑↑↓	↑↓↑	010	2
2	010	↑↓↑	↓↑↑	100	4
3	011	↑↓↓	↓↓↑	110	6

Although the table shows qubits, so far the calculation has been done classically and the computer would have to run the multiplication program four times to get the answers to these four calculations. However, if instead we start with a quantum superposition of all four states of the three qubits, we obtain a superposition of all four answers in a single step:

Superposition of {↑↑↑, ↑↑↓, ↑↓↑, ↑↓↓} × 2 becomes
   Superposition of {↑↑↑, ↑↓↑, ↓↑↑, ↓↓↑}

Before getting carried away by our success, we must remember that the quantum superposition is not something we can observe directly and we shall have to see if it is possible to extract the answer we want from the superposition. The problem here is that we can measure only one component of spin in any experiment.

We saw that for a single qubit, in, say, a right state, the act of measuring the up/down spin would give us the answer up or down at random and would destroy the information we previously had about the state. Extending this to the present case, this means that, although we could measure, say, the up/down components of all four particles, this would give us only one of the output numbers – i.e. the result of only one of our calculations – and which one would be random and unpredictable. So what is the point of a quantum computer? The answer to this is that, in some cases, the result we are interested in is much shorter (contains far fewer bits) than the data that is being processed. Consider, for example, looking up a number in a telephone book: the data to be interrogated is the whole book, but the output is just a single phone number. Similar considerations apply when a computer search engine is used to find a webpage. If we can apply quantum computation to such a task, we might hope that its power would be used to examine, say, the whole phone book at once and then present the result we want in a form that could be interrogated by a measurement. Theoretical procedures for carrying out such searches have been developed.

Another example of this type of calculation that receives a lot of attention is the factorization of a number into its prime components. For example the number 15 equals 5 multiplied by 3. Any of us can perform the sum $5 \times 3 = 15$. If you were asked what numbers multiply together to give 15, you could find this out quite quickly – if you did not already know. But suppose you were asked to find the two numbers whose product equals 3071: you would probably have to try quite a few combinations before hitting on the answer, which is 37×83. And if I asked you for the factors of 30,406,333, you would have to spend a considerable time with a calculator before getting the answer 4219×7207. A simple program running on a PC takes about a minute to obtain the prime factors of a twenty-digit number. This time increases very rapidly as the number gets longer and it is estimated that the most powerful conventional computer we have would take many millions of years to factorize a 100-digit number. Yet, in every case you can quickly check that the factors are correct by performing a single multiplication.[4]

This problem turns out to be one that is particularly suited to a quantum computer. The details of how it could be done are complex and technical, but in essence the computer is able to test a large number of possible products simultaneously in a superposition. As we only want to know the particular product that is correct and do not want to know about the others, we need to extract only a very small part of the total information contained in the final superposition, and it is possible to achieve this by performing a single measurement. There is considerable demand for a computer that could solve this problem for large numbers in a reasonably short time. One reason for this again relates to encryption, involving a process known as 'public key cryptography'. Using this technique, the receiver (Bob) openly sends the sender (Alice) a number equal to the product of two prime factors, which only he knows. Alice uses this number to encode a message following a known procedure and sends the coded message openly to Bob. The coding process is such that, although Alice only needs to know the original product in order to create the coded message, only someone who knows its prime factors can decode it, and Bob is the only person who does. For this reason the keys used are long enough to make it difficult to know the prime factors: they typically contain several hundred digits and a conventional computer would not be expected to find the prime factors in less than a million years of computing time. However, if an eavesdropper could use a quantum computer to obtain the factors, the same calculation would be completed in a few minutes or less. Thus quantum physics could provide a powerful code-breaking tool. This would probably force those sending coded messages to abandon public keys in favour of private key cryptography, in which Alice and Bob both know the key to the code they are using and protect it from everyone else. Perhaps ironically, as we saw earlier, quantum physics can also provide a secure method for exchanging such private keys.

So why is quantum computing still a pipe dream instead of an everyday tool? The main obstacle to progress in this field is the difficulty of forming superpositions of states that are composed of more than one or two qubits. These difficulties are associated with what is known as 'decoherence' which relates to the fact that much of the information in a superposition is lost when a measurement is made. We shall return to the controversial question of what actually constitutes a quantum measurement in the next chapter, but for the moment we note that decoherence occurs not only when we measure the state of a quantum system but also whenever a particle interacts with its environment. To preserve the superposition for a significant time, we must protect the quantum system from its surroundings and this becomes harder and harder the greater the number of qubits that are involved. To date, coherence has been preserved on some systems containing up to seven qubits and these have actually been used to show that $15 = 5 \times 3$! However, we are still very far from being able to build a machine with the one hundred or so qubits needed to represent numbers big enough to be useful in cryptography.

summary

In this chapter we have considered the application of quantum physics to the processing of information. The main points discussed are:

- Classically, numbers are represented by strings of binary bits, which can have the value 1 or 0. The quantum equivalent is known as a 'qubit'.
- A qubit is a quantum system that can exist in one of two states or in a superposition of these. An example of a qubit is a particle with spin.
- Cryptography is the coding of messages to make them inaccessible to an eavesdropper. This may involve a key, which has to be known to both the sender and the receiver.
- Qubits can be used to ensure secure key exchange. Each bit is represented either by up/down spin or right/left spin at random. Any eavesdroppers can be detected because their actions inevitably change the quantum state of at least some of the qubits.
- Because qubits can exist in superpositions of 1 and 0, they can in principle be manipulated to carry out a superposition of

calculations simultaneously, although not all the results are available to an observer.

- If a quantum computer built on these principles could be built, it would be able to carry out some tasks, such as finding the factors of a large number, very much faster than any classical computer.
- The practical obstacles to the building of a useful quantum computer are great and may be insurmountable.

notes

1. This is actually true only for some types of fundamental particle, but it does apply to all those we are familiar with – the electron, proton and neutron. Some other more exotic particles have three, four or more possible spin directions, and some others have no spin at all.

2. The particles used were actually atoms of silver evaporated from a hot oven. The silver atom contains forty-seven electrons, but forty-six of these are in states containing pairs of electrons of opposite spin and hence oppositely directed magnetic moments, which cancel out. The net spin and magnetic moment of the atom is therefore just that of the remaining (forty-seventh) electron.

3. This spin rotation can be contrasted with the Stern–Gerlach experiment discussed earlier, where the effect of the magnet was to deflect the particles passing through it. The avoidance of significant spin rotation in a Stern–Gerlach magnet is an important constraint on its design.

4. We should note that this discussion is limited to numbers that have only two factors, both of which must therefore be prime numbers. We are therefore excluding other numbers such as $105 = 5 \times 3 \times 7$.

what does it all mean?

We began our discussion of quantum physics with wave–particle duality. Light, which was traditionally thought of as a form of wave motion, sometimes behaves as if it were a stream of particles, while objects, such as electrons, which had always been thought of as particles were found to have wave properties. In the earlier chapters, we avoided any detailed discussion of these concepts, and instead concentrated on explaining how they are applied to model the behaviour of atoms, nuclei, solids, etc. In this chapter we shall return to questions of principle and the conceptual problems of the subject. A word of warning: this is an area of considerable controversy, where there are a number of alternative approaches, which means that our discussion is more philosophy than physics.

We shall begin our discussion by considering the 'Copenhagen interpretation', which is the conventional view among physicists. Some alternative approaches are discussed briefly towards the end of the chapter. To assist our discussion I shall introduce another property of light, known as its 'polarization', since this provides a comparatively simple model that illustrates most of the problems of quantum physics. To understand what polarization is, we return to the classical model we first met in Chapter 2, of light as an electromagnetic wave. An electromagnetic wave is one in which an electric field varies periodically in space and time. The important point for our present purpose is that this field points along some direction in space, as illustrated in Figure 8.1. A wave in which this direction is horizontal is said to be 'horizontally polarized' (we

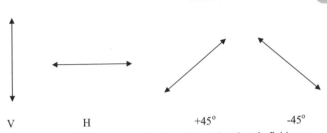

Figure 8.1 For a light wave coming towards us the electric field may oscillate vertically, horizontally or at some angle in between, but the oscillation is always perpendicular to the direction of the light beam.

shall refer to this as 'H') whereas one with the field varying vertically is 'vertically polarized' ('V'). There is nothing special about these directions: waves can also be polarized in any intermediate direction, such as at 45° to the horizontal. The polarization of much of the light we encounter from day to day, such as daylight or the light from tungsten or fluorescent lights, is not well defined because it is continually changing. To create a beam of polarized light, we must pass it through a 'polarizer'.

An example of a polarizer is the Polaroid used in the lenses of sunglasses: when randomly polarized light passes through this material, half of it is absorbed while the other half passes through with a definite polarization that is defined by the orientation of the lens. Thus the intensity of the light is halved, while the colour balance is unchanged because light of all colours is treated the same, which is why Polaroid is so suitable for sunglasses. A less familiar form of polarizer is a crystal of the mineral calcite: when unpolarized light passes through this device, it is split into two beams, one of which is polarized parallel to a particular direction defined by the crystal, while the other is perpendicular to it. Unlike Polaroid, where half the light is lost, all the light emerges in one or other of these beams. It is important to note that a calcite crystal is not like a filter letting through only the small amount of light that was already polarized in the correct direction. Instead it divides or 'resolves' the light into two components with perpendicular polarization and the sum of their intensity equals that of the incident beam, whatever its initial polarization – no light is lost. The details of how all this works are not relevant to our purpose

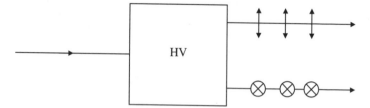

Figure 8.2 In discussing polarization, we represent a polarizer as a box with a legend indicating the direction of the polarization axis. In the example shown, the box resolves the incident light into components polarized in the vertical and horizontal directions.

and we will represent a polarizer such as a calcite crystal as a box where a beam of light enters from one side and emerges as two beams with perpendicular polarization from the other, as in Figure 8.2.

Polarization is a property of an electromagnetic wave, but does it have any relevance to the particle model of light? We could test this by passing very weak light through a polarizer set up like that in Figure 8.2: we would find photons (the particles of light first mentioned in Chapter 2) emerging at random through the two output channels, corresponding to horizontal (H) and vertical (V) polarization, respectively. To confirm that the photons really can be considered as having the property of polarization, we could pass each beam separately through other polarizers also oriented to measure HV polarization. We would find that all the photons emerging from the H channel of the first polarizer would emerge from the H channel of the second one – and similarly for V. This gives us an operational definition of photon polarization: whatever this property may really be, we can say that horizontally and vertically polarized photons are those that emerge from the H and V channels, respectively, of a polarizer. Thus, the properties of polarized photons are in many ways similar to those of spinning electrons discussed in Chapter 7 and a polarized photon is another example of a qubit.

Now suppose that photons that are neither horizontally nor vertically polarized, but polarized at 45° to the horizontal, are incident on an HV polarizer, as in Figure 8.3. If we were to carry out

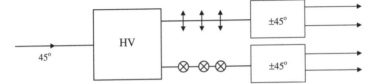

Figure 8.3 45° polarized photons incident on an HV polarizer emerge as either horizontally or vertically polarized. They are then passed through ±45° polarizers from which each emerges from one or other of the two channels at random. Thus, they appear to have lost any memory of their original polarization. We conclude that polarization measurements generally change the polarization state of the measured photons.

such an experiment, we would find that half the photons would emerge from the H channel and half from the V channel at random, and from then on they would behave as if they were horizontally and vertically polarized, respectively. This illustrates several important fundamental aspects of quantum physics. First, as far as anyone knows, this is a genuinely random process: which path will be followed by any particular photon is completely unpredictable. As such, it differs from the apparent randomness of, say, tossing a coin, where the result – heads or tails – could actually be calculated in advance if we made careful enough measurements of the forces acting on the coin as it was spun. The photons approaching the box are all identical and have +45° polarization; the fact that they emerge randomly from the H and V channels reflects a fundamental randomness or lack of causality in nature. As we saw in Chapter 1, before quantum physics came along, it was generally believed that strictly causal laws of nature such as Newton's mechanics determined everything, so that all motion would be the result of the action of known forces (as in the coin-tossing example just mentioned). In the quantum world, however, this no longer holds: randomness and indeterminism are a fundamental property of nature.

A second fundamental idea illustrated by photon polarization is that in general a measurement affects and alters the state of the object being measured. Thus a photon enters an HV polarizer in, say, a 45° state and emerges with its polarization state altered to be H or V. An inevitable result of this process is that the initial 45°

state is destroyed in the sense that all information about whether its polarization was +45° or −45° is lost. We can demonstrate this by re-measuring the ±45° polarization by passing, say, the V photons through another ±45° polarizer: we will find that they now emerge at random from the +45° and −45° channels, as in Figure 8.3. The same point was made when we discussed spin in Chapter 7 (cf. Figure 7.1).

A third principle of quantum physics is that, although the individual events occur at random, the probability of their occurrence can be calculated. By this we mean that we can predict how many photons will emerge through each channel after a large number of particles have passed through. In the case of 45° photons passing through an HV polarizer, the division is fifty–fifty as we might expect from the symmetry of the situation, but if we rotate the polarization of the incident beam, to be more nearly horizontal, then the probability of the photons emerging in the H rather than the V channel increases until all of them pass through when the incident polarization is actually H. Referring to Figure 8.4, we see that polarization in any general direction can be thought of as a sum, or superposition (cf. Chapter 7 again), of H and V components. The probability of obtaining H or V as a result of a

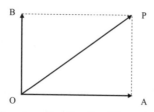

Figure 8.4 A vibration along the direction OP can be thought of as a combination or superposition of vibrations along OA and OB. If light with electric field amplitude OP (intensity OP^2) is passed through a calcite crystal whose axis points along OA, the amplitude of the horizontal and vertical components of the transmitted light will be equal to OA and OB, respectively, with corresponding intensities OA^2 and OB^2. This implies that the probability of detecting an H or V photon is proportional to the intensity of the corresponding light beam.

measurement is proportional to the square of the associated component.

These three principles – randomness of individual outcomes, alteration of state by measurement, and our ability to calculate probability – underpin the conventional interpretation of quantum physics. It is interesting to see how they connect with the basic ideas of quantum physics that we discussed in earlier chapters. Remember wave–particle duality. I said in Chapter 2 that we cannot know where a particle is until we observe it, but we can calculate the probability of finding it in a particular place from the intensity of the wave at that point. Thus, in the two-slit experiment, if we observe an interference pattern, we do not know through which slit the particle has passed, but if, instead, we look at the slits there is equal probability of finding it in one or the other. Moreover, if we detect the particles passing through the slits, we will find that they pass through one or the other at random, approximately equal numbers passing through each. However, if we do record through which slit the particle has passed, we destroy the interference pattern: the act of measuring the particle position has altered the state of the system.

The consequences of this way of thinking are even more radical than may appear so far. Consider a photon polarized in a 45° state and we ask the question 'Is this photon horizontally or vertically polarized?' But this is surely a meaningless question: the polarization is neither pointing upwards nor from side to side; it is pointing at an angle. It might make some sense to say that it is pointing partly up and down *and* partly from side to side (i.e. it is in a superposition of an H and a V state), but it is certainly not doing *either* one of these *or* the other. To ask this question is as meaningless as asking if a banana is either an apple or an orange. Thus, when we say that we 'measure' the HV polarization of a 45° photon, we are using the word in a rather different sense from the normal one. When we measure, say, the length of a piece of string, we have no problem in assuming that the string has some value of length before we put it on the ruler, but a quantum measurement is in general quite different. As we saw above, it alters the state of the system in such a way as to give reality to a quantity that was indefinable in its previous context. Now consider the implications of this way of thinking for measurements of particle position. Most of us tend to assume that a particle always has to be 'somewhere' even when it is not being observed, but this is not true in the quantum context: if a particle is in a state where its position is unknown, then to think about it even having a position is just as meaningless as ascribing H or V polarization to a particle in a 45° state. It is meaningless to say that the particle has passed through one slit or the other when an interference pattern is formed. Similarly, it is wrong to think that an electron in an atom is at any single point within it. However, just as the 45° state can be thought of as a superposition of H and V, we can think of the wave function as representing a superposition of possible positions, the contributions to the superposition from any point being weighted according to the size of the wave function at that point. Thus the particle passing through two equal-size slits is in a superposition with equal contributions from each slit, while the superposition state for an electron in a hydrogen atom has most of its contributions from a sphere of radius about 10^{-10} m centred on the nucleus.

The philosophy underlying this way of thinking was largely developed by Niels Bohr and others working in Denmark in the 1920s and 1930s and for this reason has become known as the 'Copenhagen interpretation' of quantum physics. It was heavily criticized by Albert Einstein, among others, and is not without its critics today, as we shall see later in this chapter. However, it is still the orthodox interpretation accepted by most working physicists and we shall spend some more time developing it further before explaining what some perceive to be its weaknesses and discussing some alternative approaches. Underlying Bohr's philosophy is a form of what is known as 'positivism',[1] which can be summed up in a phrase from the philosopher Wittgenstein: 'whereof we cannot speak, thereof we should remain silent'. In the present context this can be interpreted as saying that if something is unobservable (e.g. simultaneous knowledge of the HV and ±45° polarization of a photon), we should not assume that it has any reality. In some philosophical contexts this is a matter of choice – we can imagine that there are angels dancing on a pinhead if we want to – but, from the Copenhagen point of view at least, in quantum physics it is a matter of necessity. This kind of thinking is certainly counter-intuitive to anyone familiar with classical physics; for example, we find it difficult not to believe that an object must always be 'somewhere'.

When we have learned about wave–particle duality, we may well be prepared to accept this and say something like 'When the particle is not being observed it is actually a wave.' But this statement also attributes reality to something that is unobservable. We showed in Chapter 2 that the wave properties emerge when we perform experiments on a large number of particles. This is the case, for example, when an electromagnetic wave is detected by a television set or when we observe an interference experiment after many photons have arrived at the screen (even though they may have passed through the apparatus one at a time). However, if we say the wave is 'real' when we are considering an individual object, we are again suggesting that something unobservable is 'really there'. The wave function should not be interpreted as a physical wave; it is a mathematical construction, which we use to predict the probabilities of possible experimental outcomes.

Philosophers often refer to the properties of an object as its 'attributes': those of a typical classical object include permanent quantities such as its mass, charge and volume, as well as others, such as position and speed, that may change during the particle's motion. In the Copenhagen interpretation, the attributes an object possesses depend on the context in which it is being observed. Thus a photon that is observed emerging from the H channel of an HV polarizer possesses the attribute of horizontal polarization, but if it is then passed through a ±45° polarizer it loses this attribute and acquires the attribute of +45° or −45° polarization. It may be harder for us to accept that an attribute such as position has a similarly limited application, but quantum physics forces us to adopt this counter-intuitive way of thinking.

We can develop these ideas further by giving some consideration to the meaning and purpose of a scientific theory. A useful analogy is a map that might be used to navigate around a strange city. Though the map is normally much smaller than the physical area it represents, it aims to be a faithful representation of the terrain it is modelling: depictions of streets and buildings on the map are related to each other in the same way as they are in reality. Clearly, the map is not the same thing as the terrain it models and indeed is different from it in important respects: for example, it is usually of a different size from the area it represents and is typically composed of paper and ink rather than earth and stones. A scientific theory also attempts to model reality. Leaving quantum physics aside for the moment, a classical theory attempts to construct a 'map' of physical events. Consider the simple case of an object, such as an apple, being released and allowed to fall under gravity: the apple is at rest, is released, accelerates, and stops when it reaches the ground. At every stage of the motion all the relevant attributes of the object are represented in our map (e.g. the time, height and speed) are denoted by algebraic variables and to construct a map of the object's motion, we carry out some mathematical calculations. However, the real apple falls to the floor in the predicted time even though it is quite incapable of performing the simplest mathematical calculation! The aim of science is to construct the most detailed and faithful map of physical reality as is possible. This can require extensive use of mathematics, which is used to construct a map of reality, but the map is not reality itself. We also have to take care to choose a map that is appropriate to the physical situation we are addressing. Thus, a map based on Maxwell's theory of electromagnetism will be of little use to us if we are trying to understand the fall of an apple under gravity. Even if we do choose a map based on Newton's laws, it still has to use appropriate parameters: for example, it should include a representation of the effect of air resistance unless our apple is falling in a vacuum.

When we come to quantum physics, we have to accept that there is no single map of the quantum world. Rather, quantum theory provides a number of maps; which we should use in a particular situation depends on the experimental context or even on the experimental outcome, and may change as the system evolves in time. Pursuing our analogy a little further, we might say that quantum physics enables us to construct a 'map book' and that we must look up the page that is appropriate to the particular situation we are considering. Let's return to our example of a 45° photon passing through an HV polarizer as in Figure 8.3: before the photon reaches the polarizer the appropriate map is one in which a +45° polarized photon moves from left to right, whereas once it emerges from the polarizer the appropriate map is one representing the two possibilities of the photon being horizontally or vertically polarized, respectively; and when it is finally detected the appropriate map is one that describes only the actual outcome.

the measurement problem

The above may be difficult to accept, but it works, and if we apply the rules and use the map book properly, we will correctly calculate predictable outcomes of measurement: the energy levels of the hydrogen atom, the electrical properties of a semiconductor, the result of a calculation carried out by a quantum computer and so on. However, this implies that we understand what is meant by 'measurement' and this turns out to be the most difficult and controversial problem in the interpretation of quantum physics.

Consider the set-up shown in Figure 8.5. As before, a 45° photon passes through an HV polarizer, but instead of being detected, the two paths possible are brought together so that they can interfere in a manner similar to that in the two-slit experiment. Just as in that experiment, we do not know which path the photon passes along, so we cannot attribute reality to either. The consequence is that the 45° polarization is reconstructed by the addition of the H and V components – as we can demonstrate by passing the photon through another ±45° polarizer and observing that all the photons emerge in the +45° channel as in Figure 8.5. If, however, we had placed a detector in one of the paths and between the two polarizers, we would either have detected the particle or we would not, so we would know that its polarization was either H or V; it turns out that in such a case, it is impossible in practice to reconstruct the original state and the emerging photons are either H or V. We are led to the conclusion that the act of detection is an essential part of the measuring process and is responsible for placing the photon into an H or V state. This is consistent with the positivist approach outlined earlier, because in the absence of detection we do not know that the photon possesses polarization so we should not assume that it does. We appear therefore to be able to divide the quantum world from the classical world by the presence or absence of a detector in the experimental arrangement.

However, this begs the question of what is special about a detector and why we cannot treat it as a quantum object. Suppose

Figure 8.5 Light split into two components by an HV polarizer can be reunited by a second polarizer facing in the opposite direction (marked as VH). If the crystals are set up carefully so that the two paths through the apparatus are identical, the light emerging on the right has the same polarization as that incident on the left. This is also true for individual photons, a fact that is difficult to reconcile with the idea of measurement changing the photon's state of polarization (cf. Figure 2.7).

it were subject to the rules of quantum physics. To be consistent, we would have to say that the detector does not possess the attribute of having detected or of not having detected a photon until its state has been recorded by a further piece of apparatus – say a camera directed at the detector output. But if we then treat the camera as a quantum object, we have the same problem. At some point we have to make a distinction between the quantum and the classical and abandon our hope of a single fundamental theory explaining both. This impasse is the basis of a now well-known example of the application (or misapplication) of quantum physics, known as *Schrödinger's cat*. Erwin Schrödinger who was one of the pioneers of quantum physics and whose equation was referred to in Figure 8.5, suggested the following arrangement. A 45° photon passes through an HV apparatus and interacts with a detector, but the detector is now connected to a gun (or other lethal device) arranged so that when a photon is detected the cat is killed. We can then argue that if the HV attribute cannot be applied to the photon, then the attribute of detection cannot be applied to the detector, and that of life-or-death cannot be applied to the cat. The cat is neither alive nor dead: it is simultaneously alive *and* dead!

The quantum measurement problem just described is the heart of the conceptual difficulties of quantum physics and the source of the controversies surrounding it. Detectors and cats seem to be different kinds of object following different physical laws from those governing the behaviour of polarized photons. The former exist in definite states (particle detected or not; cat dead or alive) while the latter exist in superpositions until measured by something with the properties of the former. It seems that our dream of a successful single theory is not going to be realized, and that quantum objects differ from classical objects not only in degree, but also in kind.

We may therefore be led to conclude that there is a real and essential distinction between the experimental apparatus, particularly the detector, and a quantum object such as a photon or electron. In practice, this distinction seems pretty obvious: the measuring apparatus is large and made of a huge number of particles, and is nothing like an electron! However, it is quite hard to define this difference objectively and in principle: just how big does an object have to be in order to be classical? What about a water molecule composed of two hydrogen atoms and an oxygen atom, and containing ten electrons; or what about a speck of dust containing a few million atoms? It has been suggested that the laws of quantum physics are genuinely different in the case of an object made up from a large number of particles.

Philosophically, however, it would be appealing to have one theory for the physical world rather than separate theories for the quantum and classical regimes. Might there be a universal fundamental theory of the physical world that reduces to quantum physics when applied to a single particle or small number of particles and is the same as classical physics when the object we are dealing with is large enough? Physics has seen something like this before when the theory of relativity was developed. This appears to predict that objects moving close to the speed of light should follow different laws from those of Newton, but in fact the new principles apply to all objects: even when moving slowly, they are subject to the rules of relativity; it is just that the relativistic effects are very small and unnoticeable at low speeds. Perhaps the same is true in the present case, with the role of the speed replaced by, say, the number of fundamental particles in an object.

To test this hypothesis, we could try to demonstrate the wave properties of a large scale object by performing an interference experiment: if the laws of quantum physics are different at the large scale, then we should be unable to detect interference in a situation where it would be expected. The largest object to have had its wave properties demonstrated by an interference experiment akin to the Young's slit apparatus is the 'buckminster fullerene' molecule which is composed of sixty carbon atoms in a 'football' configuration. However, this does not mean that larger objects do not have wave properties, but rather that no experiment has been devised that would demonstrate them. The practical difficulties of such experiments become rapidly greater as the size of the object increases, but no experiment has yet been performed in which predicted quantum properties were not observed when they should have been.

We turn now to consider how the Copenhagen interpretation deals with this problem? Consider the following quote from Niels Bohr:

> Every atomic phenomenon is closed in the sense that its observation is based on registrations obtained by means of suitable amplification devices with irreversible functions such as, for example, permanent marks on a photographic plate caused by the penetration of the electrons into the emulsion[2].

We may conclude from this that Bohr was content to make a distinction between quantum system and classical apparatus. As we saw above, for all practical purposes, this is not difficult, and our ability to do so reliably underlies the huge success quantum physics has had. However, the Copenhagen approach goes further and denies the reality of anything other than the changes that occur in the classical apparatus: only the life or death of the cat or the 'permanent marks on a photographic plate' are real. The polarization state of the photon is an idealistic concept extrapolated from the results of our observations and no greater reality should be attributed to it. From this point of view, the function of quantum physics is to make statistical predictions about the outcome of experiments and we should not attribute any truth-value to any conclusions we may draw about the nature of the quantum system itself.

Not all physicists and philosophers are content with positivism and considerable effort has been made to develop alternative interpretations that would overcome this problem. All of these have their followers, although none has been able to command the support that would be necessary to replace the Copenhagen interpretation as the consensus view of the scientific community. We discuss some of them below.

alternative interpretations

subjectivism

One reaction to the quantum measurement problem is to retreat into 'subjective idealism'. In doing this, we simply accept that quantum physics implies that it is impossible to give an objective account of physical reality. The only thing we know that must be real is our personal subjective experience: the counter may both fire and not fire, the cat may be both alive and dead, but when the information reaches my mind through my brain I certainly know which has really occurred. Quantum physics may apply to photons, counters and cats, but it does not apply to you or me! Of course, I do not know that the states of your mind are real either, so I am in danger of relapsing into 'solipsism', wherein only I and my mind have any reality. Philosophers have long argued about whether they could prove the existence of an external physical world, but the aim of science is not to answer this question but rather to provide a consistent account of any objective world that does exist. It would be ironic if quantum physics were to finally destroy this mission. Most of us would much rather search for an alternative way forward.

hidden variables

An interpretation that rejects the positivism of Bohr in favour of realism (or 'naïve realism' as some of its detractors prefer) is based on what are known as 'hidden variables', by which is meant that a quantum object actually does possess attributes, even when these cannot be observed. The leading theory of this kind is known as the 'de-Broglie–Bohm model' (DBB) after Louis de Broglie, the first person to postulate matter waves, and David Bohm, who developed and expanded these ideas in the 1950s and 1960s. In DBB theory, both the particle position and the wave are assumed to be real attributes of a particle at all times. The wave evolves according to the laws of quantum physics and the particles are guided both by the wave and the classical forces acting on it. The path followed by any particular particle is then completely deter-mined and there is no uncertainty at this level. However, different particles arrive at different places depending on where they start from, and the theory ensures that the numbers arriving at different points are consistent with the probabilities predicted by quantum physics. As an example, consider the two-slit experiment: according to DBB theory the form of the wave is determined by the shape, size and position of the slits, and the particles are guided by the wave so that most of them end up in positions where the interference pattern has high intensity, while none arrives at the points where the wave is zero.

As we have noted before, the emergence of apparently random, statistical outcomes from the behaviour of deterministic systems is quite familiar in a classical context. For example, if we toss a large number of coins, we will find that close to half of them come down heads while the rest show tails, even though the behaviour of any individual coin is controlled by the forces acting on it and the initial spin imparted when it is tossed. Similarly, the behaviour of the atoms in a gas can be analysed statistically, even when the motion of its individual atoms and the collisions between them are controlled by classical mechanical laws.

De-Broglie–Bohm theory therefore reproduces all the results of conventional quantum physics without positivist baggage. Why, one might ask, is it not universally adopted? In the nineteenth century, the fact that the properties of a gas could be predicted from the statistics of the motion of its atoms was considered to be strong evidence for the existence of the atoms themselves. Some scientists (notably Ernst Mach) did not accept that this was sufficient reason to accept the existence of atoms and the issue was not finally settled until Einstein showed in 1905 that the phenomenon of Brownian motion resulted from the motion of atoms. Why, then, should we not favour the realist approach in the quantum case? One reason is that problems arise when we examine the implications of DBB theory in more detail. Some of this analysis is quite technical, but it turns out that many of the properties, such as mass and charge, that are normally thought of as particle properties are actually associated with the wave in DBB theory: the particle has the attribute of position, but little, if anything, else.

The principle objection to DBB theory is that it is what is known as a 'non-local' theory. To understand what this means, we have to consider some properties of systems containing more than one particle. Particles may exert forces (e.g. electrical or gravitational) on each other, but these forces are subject to an important constraint from the theory of relativity, which is that no influence can travel between the particles at a speed greater than that of light. As a result, if one particle changes its position this will not have any effect on the other until a short time later – i.e. at least as long as it would take light to travel between them. In the DBB version of quantum theory, however, the influence exerted by the wave on the particle is subject to no such constraint: in many cases the quantum predictions are reproduced only if it is assumed that the one particle is influenced by the properties of the wave at the position of another at the same time as it senses the influence of the wave at its own position. To reproduce quantum mechanics, DBB must not only assume the existence of hidden attributes such as particle position, but also that these are not subject to the extremely fundamental principles of physics discovered by Einstein.

We can develop this point further by considering the behaviour of a system consisting of pairs of photons emitted from a particular kind of light source; this has the property that the polarizations of the two photons in each pair are always perpendicular to each other. We mean by this that if we measure the HV polarization of one photon we will find it to be either H or V at random, as before, but the other photon of the pair will always have polarization V or H, respectively, as illustrated in Figure 8.6. This may not surprise us, but perhaps it should. Remember what happens in a polarization measurement: whatever the previous polarization (unless it happened to be precisely H or V), we expect a photon to emerge from the H and V channels at random and we

Figure 8.6 In some circumstances atoms can be made to emit a pair of photons in rapid succession. The two members of each pair move away from the source in different directions. In the set-up shown, the light source is in the centre; the right-hand apparatus measures the HV polarization of one of the two photons while that of the other is measured on the left. Whenever a right-hand photon is recorded as horizontally polarized that on the left is found to be vertical and vice versa.

concluded from this that the act of measurement has changed the polarization state of the photon. But if, as in the present case, the polarizations of the two photons are always found to be at right angles, presumably each must 'know' what is happening to the other, so how can the outcome be random? This is reinforced by the fact that if we turn the polarizers round so that each measures ±45° polarization, then whenever the right-hand photon is found to be +45° the left hand one is −45° and vice versa. The photon passing through one polarizer seems to know what measurement is being made on the other and what the result is.

One way of apparently resolving this question is to revise our belief of what happens in a polarization measurement: if the result is not actually random after all, but determined by some kind of hidden variable, then the results of the measurements on both photons would be determined in advance and there would be no problem ensuring that the results always corresponded to perpendicular polarizations. The experiment would then be like a classical one in which one person (Alice) is given a black or white ball at random, while the other (Bob) is always given one of the opposite colour. Alice and Bob separate and the colours of their balls are 'measured'. Each measurement gives black or white at random, but the two balls are always found to have opposite colours.

An experiment illustrating the same principle, though not involving polarization measurements, was suggested by Einstein and co-workers[3] in 1935. They concluded,

> If, without in any way disturbing the system, we can predict with certainty (i.e. with probability equal to unity) the value of a physical quantity, then there exists an element of physical reality corresponding to this physical quantity.

By 'element of reality' they meant something like a hidden variable that determines the value of an attribute such as polarization before this is measured, thus accounting for the correlations between the results of separate measurements in the way discussed above.

This question was again addressed by John Bell in the 1960s. He was attracted to the idea of hidden variables and disliked the conventional view that only measurable attributes of a quantum system can be treated as real. However, his principal contribution to the field was to show that the no local hidden variable model – by which is meant any model that excludes instantaneous communication between the individual photons – could be consistent with the predictions of quantum physics. This result is known as 'Bell's theorem' and it relates to a generalization of the experiment discussed above in which the two polarizers are oriented so as to measure different polarizations. Thus one might measure HV on the right-hand photon while ±45° was measured on the left. The quantum probabilities of the different outcomes can be calculated straightforwardly, but John Bell was able to show, by an argument too technical to go into here, that it was impossible for any local hidden variable theory to reproduce these. The only way round this would be for the hidden variables to be non-local – i.e. the hidden variable associated with one photon would have to know what was happening to the one associated with the other. Moreover, this communication would have to be instantaneous rather than propagating at a speed less than that of light – hence the clash with relativity mentioned above.

This result caused considerable interest and led a number of scientists to perform experiments to test whether the quantum physics predictions were indeed correct for pairs of particles or if Bell's theorem would hold. The results of all such experiments performed over the last thirty or so years have upheld quantum theory and produced results that are inconsistent with any theory based on local hidden variables.

How then does conventional quantum theory treat a situation such as the measurement of the polarization of photon pairs? Shortly after Einstein's paper came out, Bohr published a response, the key phrase of which was 'There is essentially *the question of an influence on the very conditions that define the possible types of prediction regarding the future behaviour of the system*' (Bohr's italics). Applying this to the two-photon case, Bohr is saying that if we alter the orientation of one of the polarizers, we are not affecting the photons physically, but are only changing the attributes (i.e. the allowed values of the polarization) that we can assign to the system. Returning to our map-book analogy, we must turn to a different page to find the appropriate map to describe the changed situation; this does not have a direct effect on the quantum system, but only on the language we use to describe it. Whether or not we find this satisfactory depends strongly on our own ideas and prejudices. It certainly did not satisfy Einstein, whose reaction was that Bohr's position was logically possible, but 'so very contrary to my scientific instinct that I cannot forego my search for a more complete conception'. No such 'complete conception' has yet emerged to command a consensus in the scientific community.

many worlds

We discussed earlier how the measurement problem arises because a literal application of quantum physics results in not only the photon but also the measuring apparatus being put into a superposition state, so that in the case of Schrödinger's cat we have a cat that is both alive and dead. It turns out that one way to avoid this problem is to ignore it. Suspending disbelief, let us see what happens if we take the above scenario seriously and ask how we could tell that the cat really was in such a state. The reason we know that a particle passing through a two-slit apparatus is in a superposition of being in one slit and being in the other is that we can create and observe an interference pattern. However, to do the equivalent thing with the cat, we would have to bring the wave function representing all the electrons and atoms in both live and dead cat together to form an immensely complicated interference pattern. In practice, this is a completely impossible task. We might think that all we would need to do to show that the cat is in a live/dead superposition is to look at it, but this is not the case. If we treat ourselves as part of the quantum world, this action puts us into a superposition of states in one of which we see that the cat is dead and in the other of which we see it alive. (We mentioned this earlier when discussing subjectivism and rejected it out of hand.) However, it is just as impossibly difficult to do an interference experiment on ourselves as it is on a cat, so we could not in practice know that we were in a superposition state. Detailed quantum calculations show that it would be impossible for the 'me' that is in one half of the superposition to be aware of the 'other me' in the other. This means that the whole system has 'branched' and there is no way that an observer on the branch containing the dead cat can ever know anything about the existence of the one who sees the cat alive.

We can therefore resolve the measurement problem by ignoring it, but only at the cost of accepting the existence of branches containing copies of ourselves and of our cats, which can never observe each other. Moreover, the branching does not stop with the observer but extends to everything with which the system or the observer interacts. Hence the term 'many worlds' or its alternative, 'branching universe': everything branches and this does not happen only when someone sets up a polarization-measuring apparatus with or without a cat. Similar types of processes are happening all the time in the physical world, which would by now have created an unimaginably large number of branches. At least the fact that this can be taken seriously shows that not all scientists are positivists!

The advantages of the many-worlds approach are that it preserves realism, albeit of all the branches rather than just the one we know, and that nothing has to be added to quantum theory to deal with the measurement problem. One of its disadvantages is the extravagance involved in the huge number of branches. For these reasons, many-worlds theory has been described as 'cheap on postulates, but expensive on universes'. However, there is another difficulty, which is the problem of defining probabilities. When a 45° photon passes through an HV apparatus, we say that there is a fifty per cent probability of it emerging in, say, the H channel, which implies that there is a also a fifty per cent probability of it *not* doing so. But this is inconsistent with the fact that in a many-worlds scenario everything happens. Probability implies a 'disjunction' – something happens or it doesn't; it surely ought not be applied to a 'conjunction' where both outcomes exist. Supporters of many-worlds theory have suggested ways to overcome this problem, but despite this the model has not won a consensus in the scientific community.

summary

This chapter has discussed the concepts and problems associated with understanding quantum physics. The main points are:

- Our discussion used the example of photon polarization. If light passes through an analyser such as a calcite crystal, it is resolved into two components whose directions are either parallel or perpendicular to a direction defined by the crystal.

- The outcome of a photon polarization measurement is in general unpredictable, though the relative probabilities of the possible outcomes can be calculated. The act of measuring a photon's polarization generally destroys previous information about its polarization.

- From the Copenhagen point of view, unobservable attributes, such as the HV polarization of a photon known to be in a +45° state, have no reality.

- The measurement problem arises when we apply the super-position principle to objects such as photon detectors, Schrödinger's cat or even ourselves: in some circumstances, they also appear to be in superposition states whose attributes have no reality.

- Alternative approaches to the quantum measurement problem include subjectivism, hidden variables and many worlds.

- Subjective theories postulate that superpositions collapse only when the information enters a human, conscious mind.

- Hidden variable theories postulate the reality of at least some unobservable attributes. Bell's theorem and associated experiments have shown that such theories can succeed only if they are non-local theories inconsistent with the principles of relativity.

- Many-worlds theories accept the reality of superpositions at all levels, including ourselves. They lead to the concept of large numbers of alternative parallel universes, unaware of each other's existence. A problem with many-worlds theories, apart from the extravagance of their postulates, is the difficulty of defining probabilities in a context where everything happens.

notes

1.   It should be noted, however, that there was little if any connection between Bohr's ideas and the developments in positivist philosophy that were going on around the same time.

2.   N. Bohr, *Atomic Physics and Human Knowledge*. New York, Wiley, 1958.

3.   Boris Podolski and Nathan Rosen.

conclusions

The twentieth century could well be called the era of the quantum. One hundred years on from Einstein's realization that light consists of quanta of fixed energy, how far have we come and where may we be going? This chapter attempts to gather together some strands from the earlier chapters, to place these in historical context and to make some guesses about what might be in store for the twenty-first century.

the early years

Progress was quite slow for the first twenty years or so after Einstein's explanation of the photoelectric effect in 1905. However, once the principle of wave–particle duality and its mathematical development in the Schrödinger equation were established they were quickly applied to elucidate the structure of the atom and its energy levels (Chapter 2). Within another twenty years, quantum physics had been successfully applied to a wide range of physical phenomena, including the electrical properties of solids (Chapter 4) and the basic properties of the atomic nucleus. The possibility of nuclear fission (Chapter 3) was understood in the late 1930s and this led to the first nuclear explosion in 1945 – less than twenty years after Schrödinger first published his equation.

since 1950

The second half of the twentieth century witnessed an explosion in the development of our understanding of the principles and applications of quantum physics. One example of this was the discovery of quarks (Chapter 1), which are now part of the standard model of particle physics. This emerged from the results of experiments involving very high-energy collisions between fundamental particles, such as electrons and protons; it applied the principles of both quantum physics and relativity to address the question of the internal structure of the proton and neutron. Just as an atom or a nucleus can be excited into higher energy states, similar excitations occur when fundamental particles collide with each other at very high speeds. The products of these collisions can be thought of as excited states of the original particles and the fields associated with them, but the energy changes are so great that the associated relativistic mass change can be several times the mass of the original particle. As a result, excitations to such states are often thought of as creating new short-lived particles, which recover their original form in a very short time – typically 10^{-12} s. The construction of the machines required to carry out such fundamental experiments has involved effort and expense approaching that of the space programme.

Rather more modest, but many would say just as fundamental, have been investigations of the properties of bulk matter. The quantum explanation of superconductivity (Chapter 6) was one of the most exciting intellectual triumphs of the second half of the twentieth century. The electrical properties of many solids change radically and suddenly at sufficiently low temperatures; this results in the electrons in the solid forming a coherent quantum state that spans the whole solid and leads to the complete loss of any resistance to the flow of electric current. Superconductivity is a robust property that can be destroyed only by raising the temperature or by subjecting the material to a sufficiently high magnetic field. As we have seen, the technological applications of this phenomenon are already significant and their potential may be even greater. Other large-scale quantum phenomena have also emerged during this period, though they are less well known – perhaps because their potential for application is not so obvious. One example is the 'quantum Hall effect', which relates to the propeties of a thin film of semiconductor that is subject to a high magnetic field and carrying an electric current. In this situation, a voltage appears across the sample which is determined by the size of the applied field and is quantized, adopting one of a discrete set of values.

Progress in the practical application of quantum physics was also enormous during the second half of the twentieth century. The development of controlled nuclear fission (Chapter 3) led quickly to the establishment of the nuclear power industry, which in some countries now provides the majority of the nation's electricity (over seventy-five per cent in the case of France). The civil application of fusion has turned out to be a much greater challenge, but research has now brought us to the point where this may soon be a real possibility. The information revolution resulting from the development of semiconductors and the computer chip (Chapter 5) took place in the last quarter of the twentieth century and has arguably been as dramatic and important as the industrial revolution two hundred years earlier. Because of the quantum properties of silicon, we can compute at enormous speeds, communicate across the globe and beyond and download information via the worldwide web. Moreover, applying quantum physics directly to information processing (Chapter 7) has recently opened up the possibility of developing even faster and more powerful techniques in this field.

Another application of quantum physics that we have only touched on is that to chemistry and biology. We have seen (Chapter 3) some simple examples of how quantum physics underlies the formation of bonds between atoms to form molecules. Chemists have now increased their understanding of the quantum physics of chemical bonding. This has contributed to the discovery and creation of a huge variety of molecules, including many medical drugs and the plastics used to construct everything from everyday kitchen utensils to the specialist materials used in spacecraft.

It is arguable how much recent progress has been made in understanding the conceptual basis of quantum physics (Chapter 8). Bohr developed the Copenhagen interpretation in the 1920s and 30s and this is still the orthodoxy accepted by the majority of working physicists. Considerable effort has gone into the construction of alternatives – particularly hidden-variable theories and the many-worlds interpretation – during the last fifty years or so. Much of the research on hidden-variables theories, by both supporters (such as John Bell) and skeptics, has diminished rather than increased their credibility. In contrast, it may surprise readers to know that, despite its ontological extravagance, the many-worlds interpretation is probably the second most popular among professional physicists.

the future

As far as fundamental physics is concerned, the more powerful machines currently being built will enable the study of even higher-energy particle collisions: many expect the standard model of particle physics to break down in this regime and to be replaced by another that will produce new and exciting insights into the nature of the physical world at this level. In the area of condensed matter, investigations into the behaviour of matter at extremes of temperature and field will continue and may well throw up new and fundamental manifestations of quantum physics.

Without a reliable crystal ball, it would be perilous to predict future applications of quantum physics. We can certainly expect conventional computers to continue increasing in power and speed for some years to come: the ability of silicon to surprise should never be under-rated. The study of superconductivity will certainly continue, but unless and until malleable materials appear that remain superconducting up to room temperature, only quite specialist applications seem likely. Currently, a huge effort is being devoted to the development of devices to perform quantum computations (Chapter 7). Whether this will succeed within the foreseeable future is difficult to judge; anyone thinking of betting on this happening would be well advised to exercise considerable caution.

Hopefully, the dangers of the continued burning of fossil fuels will be better appreciated very soon and the pressure to develop alternatives will increase. This may well result in the development of a new generation of nuclear reactors as well as improvements in green technologies, including those relying on quantum physics, such as the photovoltaic cell (Chapter 5). The problem is so serious that we would do well to abandon arguments about the advantages and disadvantages of the different alternatives: almost certainly, all possible approaches will have to be exploited if we are to avoid a major catastrophe within the next fifty to one hundred years.

It seems unlikely that the philosophical questions associated with quantum physics (Chapter 8) will be resolved in the near future. In this regard, quantum physics appears to be a victim of its own success. The fact that it has provided successful explanations for such a huge range of physical phenomena, and that it has not so far failed, means that the debate is about alternative interpretations rather than any need for new theories. So far at least, any new way of looking at quantum phenomena that predicts results different from those of standard quantum physics has been proved to be wrong. Some new theory in the future may break this pattern and, if it did, this would likely be the most exciting fundamental development since the invention of quantum physics itself. Perhaps such a development will emerge from the study of the quantum properties of black holes and the big bang that created our universe. New theories will almost certainly be needed in this area, but it is by no means obvious that these will also address fundamental questions such as the measurement problem. The philosophical debate seems likely to continue for a long time to come.

I hope the reader who has got this far has enjoyed the journey. I hope you agree that quantum physics does not need to be rocket science and that you now understand why some of us have devoted a considerable part of our lives to trying to understand and appreciate what is arguably the greatest intellectual achievement of the human race.

glossary

Words in *italics* are listed elsewhere in the glossary.

Absolute zero of temperature The temperature at which all thermal motion ceases; it is equivalent to −273 degrees on the Celsius scale.

Acceptor level A set of empty states created when impurity atoms that contain one *electron* per atom more than the host atoms are added to a *semiconductor*. They lie just above the *valence band* and can capture *electrons* from it to form *holes*.

Alpha particle Two *protons* and two *neutrons* bound together; it forms the *nucleus* of the helium atom.

Amplitude The maximum displacement of a wave.

Atom The building blocks of all matter. An *atom* contains a number of *electrons* and a *nucleus* that carries a positive charge equal and opposite to the total charge on the *electrons*.

Base The central semiconducting layer of a *transistor*.

Bell's theorem A mathematical proof that any hidden-variable theory whose predictions agree with those of quantum physics must be non-local.

Binary A number system in which numbers are expressed as powers of two, in contrast to the decimal system in which powers of ten are used.

Binary bit A quantity that can adopt the values 1 or 0 and which is used to express numbers in *binary* form.

Brownian motion The irregular motion of pollen grains suspended in a liquid, caused by the random motion of the *atoms* in the liquid.

Chain reaction A series of *fission* events that occur when the *neutrons* emitted from a *nucleus* undergoing *fission* trigger the *fission* of other *nuclei*.

Classical The set of theories used to describe physical events before the advent of quantum physics.

Closed shell A set of energy states of similar energy in an *atom* which are all occupied by *electrons*.

Collector The semiconducting layer in a *transistor* which collects charged carriers from the *base*.

Conduction band A partly filled band of energy levels in a metal or *semiconductor*. The *electrons* in the conduction band are mobile and can carry electric current.

Conservation of energy The principle that energy cannot be created or destroyed, but only converted from one form to another.

Cooper pair A pair of *electrons* of opposite *momentum* that are bound together in a *superconductor*.

Copenhagen interpretation The standard interpretation of quantum physics, which denies the reality of unobservable attributes.

Coulomb The standard unit of electric charge; also used as an adjective to describe electrostatic interactions and fields.

Critical current The largest current that can flow through a *superconductor* or *Josephson junction* without destroying the *superconductivity*.

De Broglie relation The rule that the wavelength of a matter wave equals *Planck's constant* divided by the particle momentum.

Deuterium An *isotope* of hydrogen whose *nucleus* contains one *proton* and one *neutron*.

Donor level The set of filled states created when impurity *atoms* that contain one *electron* per *atom* more than the host *atoms* are added to a *semiconductor*. They lie just below the empty *conduction band* and can donate *electrons* to it.

Electromagnetic radiation Waves constructed from oscillating electric and magnetic fields propagating through space. Examples include light waves and radio waves.

Electron A fundamental point particle that carries a negative charge.

Emitter The semiconducting layer of a *transistor* that emits charged carriers into the *base*.

Energy gap A band of energies in a metal or *semiconductor* which normally contains no states for *electrons* to occupy.

Excited state Any quantized energy state other than the *ground state*.

Fermi energy The energy of the highest filled energy level in a metal.

Fission A process where a *nucleus* splits into fragments, releasing energy along with *neutrons*.

Flux quantum When a magnetic field passes through a loop of *superconductor*, the total field through the loop (the flux) always equals a whole number of flux quanta.

Fossil fuel A fuel such as coal or natural gas.

Free electrons *Electrons* in a metal that are not bound to individual *atoms*.

Fusion A process where two *nuclei* join together with the release of energy.

Global warming An increase in the overall temperature of the Earth's atmosphere.

Greenhouse effect Light passing through the glass of a greenhouse warms its contents, which radiate heat, but this cannot escape through the glass. A similar effect occurs in the Earth's atmosphere owing to the presence of gases such as carbon dioxide.

Ground state The lowest energy state of a quantum system such as an *atom*.

Hidden variables Quantities that are real though unobservable and are postulated to produce a realistic interpretation of quantum physics.

High-temperature *superconductors* Materials that remain super-conducting at temperatures well above twenty *kelvin*.

Hole A positive charge carrier in a *semiconductor*, created when an *electron* is removed from a full or nearly full band.

Insulators Materials that do not allow the flow of electric current.

Interference The result of the combination of two waves that reach a point by following different paths.

Ions *Atoms* that are positively or negatively charged owing to the removal or addition, respectively, of one or more *electrons*.

Isotope One of the possible *nuclei* associated with an element. Different *isotopes* of the same element have the same number of *protons* but different numbers of *neutrons*.

Josephson junction A device consisting of two pieces of *superconductor* separated by a thin insulating layer, through which a current can pass without resistance.

Joule The standard unit of energy.

Kelvin The standard unit of temperature when measured from the *absolute zero of temperature*.

Kilogram The standard unit of *mass*.

Kinetic energy Energy associated with a particle's motion. It equals half the product of the particle *mass* and the square of its speed.

Many-worlds interpretation An interpretation of quantum measurement in which different outcomes coexist in parallel non-interacting universes.

Mass A measure of the quantity of matter in a body.

Momentum The product of a particle's *mass* and its *velocity*.

Neutron An uncharged particle of similar mass to a *proton* which is a constituent of most *nuclei*.

Non-local interaction An interaction that passes between two systems instantaneously rather than at the speed of light or slower.

N-type A *semiconductor* whose charge carriers are predominately negative *electrons*.

Nucleon A name for a particle that is either a *proton* or a *neutron*.

Nucleus An object consisting of *protons* and *neutrons* tightly bound together that carries most of the *mass* of the *atom* but occupies only a small part of its volume.

Ohm's law The rule that the electric current passing through an electric circuit is the product of the applied voltage and the circuit resistance.

One-dimensional A model system in which all motion is along a line.

Photoelectric effect The emission of *electrons* from a metal when a light shines on it.

Photon A particle that carries the quantum of energy in a beam of light or other *electromagnetic radiation*.

Photovoltaic cell A device that directly converts light energy into electrical energy.

Planck's constant A fundamental constant of nature which is involved in determining the size of quantized quantities.

P-n junction A connection between a *p-type* and an *n-type semiconductor* which allows current flow in one direction only.

Polarization The direction of the electric field associated with an electromagnetic wave.

Potential energy The energy associated with a field, such as a gravitational or electric field.

Proton A particle that carries a positive charge equal and opposite to that on an *electron* and has a *mass* about two thousand times the *electron mass*.

P-type A *semiconductor* whose charge carriers are predominately positive *holes*.

Quantum computing The application of the principles of quantum physics to perform some types of calculation very much faster than is possible with a *classical* computer.

Quantum cryptography The application of the principles of quantum physics to the encoding of information.

Quantum mechanical tunnelling A process whereby wave–particle duality allows a particle to pass through a barrier that would be impenetrable *classically*.

Qubit A quantum object that can exist in either one of two states or in a *superposition* made up from them.

Resistor A device that resists the flow of electric current around a circuit and is subject to *Ohm's law*.

Schrödinger's cat A name for a scenario in which the rules of quantum physics appear to predict that a cat can be placed in a *superposition* of a live and a dead state.

Schrödinger equation The fundamental equation used to calculate the form of the *wave function* in quantum physics.

Screening The property whereby a piece of metal prevents an electric field from penetrating it.

Semiconductor A material with an electronic structure similar to that of an *insulator* but with a small *energy gap*.

Spin A property of *electrons* and other fundamental particles whereby they behave as if they were rotating about an axis. Unlike *classical* rotation, spin always has the same magnitude and is either parallel or anti-parallel to the direction of measurement.

SQUID A 'superconducting quantum interference device'. It consists of a circuit containing two *Josephson junctions* and can be used to make very accurate measurements of magnetic field.

Standing waves Waves confined to a region of space where they cannot travel.

Superconductor A material that offers no resistance to the flow of electrical current.

Superposition A quantum state that can be considered to be composed of two or more other states.

Transistor A device composed of three pieces of *semiconductor* (*emitter*, *base* and *collector*) in contact. The size of the current flowing from the *emitter* to the *collector* is controlled by that injected into the *base*.

Travelling waves Waves free to travel through space. They do so at a given speed that depends on the nature of the wave.

Uncertainty principle A property of quantum systems whereby properties such as position and *momentum* cannot be precisely measured at the same time.

Unit cell The basic building block of a crystal, constructed from a (usually small) number of *atoms*.

Valence band A normally filled band of energy levels in a metal or *semiconductor*. If electrons are removed from the valence band, *holes* are created that can carry electric current.

Vector A quantity (such as *velocity*, force or *momentum*) that acts in some particular direction.

Velocity The speed of an object in a given direction.

Voltage A property of a battery or similar device that drives a current around an electric circuit.

Wave function A mathematical function, similar to a wave, associated with the quantum properties of a particle. The square of the wave function at any point equals the probability of finding the particle there.

Wavelength The repeat distance of a wave.

Wave–particle duality A property of quantum systems whereby their properties combine those of a *classical* particle and a *classical* wave.

index